Introduction to Urban Water Distribution

Second Edition

Volume 1: Theory

T0074634

IHE Delft Lecture Note Series

ISSN: 1567-7052

Introduction to Urban Water Distribution

Theory

Second Edition

Nemanja Trifunović

CRC Press
Taylor & Francis Group
Boca Raton London New York

CRC Press is an imprint of the
Taylor & Francis Group, an **informa** business

United Nations
Educational, Scientific and
Cultural Organization

Institute for
Water Education
in partnership with UNESCO

Second edition published 2020

CRC Press/Balkema is an imprint of the Taylor & Francis Group, an informa business

© 2020 Taylor & Francis Group, London, UK

Typeset by Apex CoVantage, LLC

Library of Congress Cataloging-in-Publication Data
Applied for

Published by: CRC Press/Balkema
                Schipholweg 107C, 2316 XC Leiden, The Netherlands
                e-mail: Pub.NL@taylorandfrancis.com
                www.crcpress.com – www.taylorandfrancis.com

Volume 1

ISBN: 978-0-367-50301-7 (Hbk)
ISBN: 978-0-367-50445-8 (Pbk)
ISBN: 978-1-003-04985-2 (eBook)
DOI: https://doi.org/10.1201/9781003049852

Volume 2

ISBN: 978-0-367-50302-4 (Hbk)
ISBN: 978-0-367-50448-9 (Pbk)
ISBN: 978-1-003-04986-9 (eBook)
DOI: https://doi.org/10.1201/9781003049869

Two-volume set

ISBN: 978-0-367-50295-9 (Hbk)
ISBN: 978-0-367-50443-4 (Pbk)
ISBN: 978-1-003-04941-8 (eBook)

Climbing is slow. Falling is fast.
The faster the climb, the longer the fall.

Contents

Preface to the second edition

In 2016, precisely ten years after the book 'Introduction to Urban Water Distribution' was first published by Taylor & Francis, I decided to start work on a second edition. I was inspired by numerous positive reactions from my students and peers, a few of whom started using my materials in their educational programmes. However, some of the contents gradually became outdated and an update with newer information became inevitable.

Despite a lot of enthusiasm for this ambition, it took me nearly three years to complete the work. This was because I took the opportunity to strengthen some sections in the book with better elaborated principles, further expand the contents with missing developments in the field, and also add new exercises tested in the classrooms during the last 10-15 years. The result is this material which grew from approximately 500 to 800 pages, with over 700 figures comprising various diagrams, drawings, and computer layouts, including some 200 photos largely from my own collection made during student fieldtrips, project missions, conference exhibitions, or simply by observing interesting water distribution practices while biking on a sunny day around my area. The practical part of the book covered in the appendices has been nearly doubled, by adding two more computer exercises with detailed tutorials, and the selection of 15 solved examination problems and true-false tests, given to the students in Delft in the period 2004 to 2019. Consequently, the electronic materials accompanying this book have also been upgraded with additional spreadsheet and computer modelling applications discussed in the workshop problems and exercises.

Following significant expansion, this book has been published in two volumes: Volume 1 covering the theory (referred to in the table of contents as chapters 1 to 6), and Volume 2 covering the workshop problems and computer modelling exercises (referred to in the table of contents as appendices 1 to 9). Although each of the volumes can be studied separately, in many places in the chapters there are references to the appendices, as well as some clarifications in the appendices that contain references to the chapters/sections. Moreover, all electronic materials have been attached to Volume 2, some of which are also mentioned in Volume 1. Hence, studying both volumes in parallel is the obvious and most effective approach.

This book is used in several water distribution-related specialist modules of the Master of Science programme in Urban Water and Sanitation at IHE Delft Institute for Water Education, in Delft, the Netherlands. In addition, it is the core material in the online version of the short course 'Water Transport and Distribution' and is also used in various training programmes in capacity development projects of IHE Delft. Participants in all these programmes are professionals with various backgrounds and experience, mostly engineers, working in the water supply sector from over forty, predominantly developing, countries from all parts of the world.

This current version is the summary of 25 years of development of IHE educational materials now put at the disposal of both students and teachers in the field of water transport and distribution. On that long road I learned continuously from consultations with experts from the Dutch water sector acting as guest lecturers in our institute and also helping us to organise field visits for our students. A few of the most important water companies in that respect are WATERNET Amsterdam, EVIDES Rotterdam, WMD Assen, and also KWR research institute in Nieuwegein. Furthermore, in my IHE career I have had numerous opportunities to interact with the participants of my training programmes conducted abroad, who have brought to my attention many applications that differ from Dutch and European practice. All these ingredients have helped me tremendously to arrive at the result that will hopefully satisfy the target audience.

Nemanja Trifunović

Introduction

This book was written with the idea of elaborating general principles and practices in water transport and distribution in a straightforward way. Most of its readers are expected to be those who know little or nothing about the subject. However, experts dealing with advanced problems can also use it as a refresher of their knowledge, while the lecturers in this field may wish to use some of the content in their educational programmes.

The general focus in the book is on understanding the hydraulics of distribution networks, which has become increasingly relevant since the large-scale introduction of computers and the exponential growth of computer model applications, also in developing countries. This core is handled in Chapter 3 which discusses the basic hydraulics of pressurised flow, and Chapter 4 which talks about the principles of hydraulic design and computer modelling applied in water transport and distribution. Exercises and tutorials resulting from these chapters are given in appendices 1 to 4.

The main purpose of the exercises is to develop a temporal and spatial perception of the main hydraulic parameters in the system for a given layout and demand scenarios. The workshop problems in Appendix 1 are a collection of calculi tackling various supply schemes and network configurations in a vertical cross-section. Manual calculation is advised here, whilst the spreadsheet lessons illustrated in Appendix 7 can help in checking the results and generating new problems. On the other hand, the tutorials in appendices 2 to 4 discuss, step by step, a computer-aided network design and renovation looking at the network layout in a plan i.e. from a horizontal perspective. Each of these exercises has been formulated with individual data sets that allows attempts with many different source/terrain configurations and demand scenarios. To facilitate the calculation process, the EPANET software of the US Environmental Protection Agency has been used as a network modelling tool. This programme has become popular amongst researchers and practitioners worldwide, owing to its excellent features, simplicity and free distribution via the Internet.

Furthermore, the book contains a relatively detailed discussion on water demand (Chapter 2), which is a fundamental element of any network analysis, and chapters on network construction (Chapter 5) and operation and maintenance (Chapter 6).

Complementary to these contents, more on the maintenance programmes and management issues in water distribution is taught in the Water Governance programmes at IHE Delft. Furthermore, the separate subjects on geographical information systems, water quality, and transient flows, all with appropriate lecture notes, make an integral part of the six-week programme on water transport and distribution, which explains the absence of these topics from the scope of this book.

The book comes with a selection of electronic materials containing the spreadsheet hydraulic lessons, a copy of the EPANET software (Version 2.12) and the entire batch of computer model input files and spreadsheet applications mentioned in appendices 1 to 4. Hence, studying with a PC will certainly help to master the contents faster.

The author and IHE Delft are not responsible and assume no liability whatsoever for any results or any use made of the results obtained based on the contents of this book, including the accompanying electronic materials. However, any notification of possible errors or suggestion for improvement will be highly appreciated. Furthermore, any equipment shown in photographs is for illustrative purposes only and is not endorsed or recommended by IHE Delft.

Chapter 1

Water transport and distribution systems

1.1 Introduction

Everybody understands the importance of water to our lives; clean water has been a matter of human concern for thousands of years. It is a well-known fact that all the major early civilisations regarded an organised water supply as an essential requisite of any sizeable urban settlement. Amongst the oldest, archaeological evidence on the island of Crete in Greece proves the existence of water transport systems as early as 3500 years ago, while the example of pipes in Anatolia in Turkey points to water supply systems approximately 3000 years old (Mays, 2000).

The remains of probably the most remarkable and well-documented ancient water supply system exist in Rome, Italy. Sextus Julius Frontinus, the water commissioner of ancient Rome in the first century AD, describes in his documents nine aqueducts with a total length of over 420 km, which conveyed water for distances of up to 90 km to a distribution network of lead pipes ranging in size from 20 to 600 mm. These aqueducts conveyed nearly 1 million m³ of water each day, which despite large losses along their routes would have allowed the 1.2 million inhabitants of ancient Rome to enjoy as much as an estimated 500 litres per person per day (Trevor Hodge, 1992).

Nearly 2000 years later, one would expect that the situation would have improved, bearing in mind the developments of science and technology since the collapse of the Roman Empire. However, there are still many regions in the world living under water supply conditions that the ancient Romans would have considered as extremely primitive. The records on water supply coverage around the world at the turn of the twentieth century are shown in Figure 1.1. At first glance, the data presented in the diagram give the impression that the situation was not alarming. However, the next figure (Figure 1.2) on the development of water supply coverage in Asia and Africa in the preceding period 1990-2000 shows clear stagnation. This gives the impression that these two continents may be a few generations away from reaching the standards of water supply in North America and Europe.

The period that followed marks significant and coordinated efforts to increase the coverage of safe water and sanitation in the world, with the ambition to reach the United Nations Millennium Development Goals (UN MDGs). The corresponding results regarding the water supply coverage shown in Figure 1.3 reflect that improvement, more so in rural than in urban areas of the developing countries

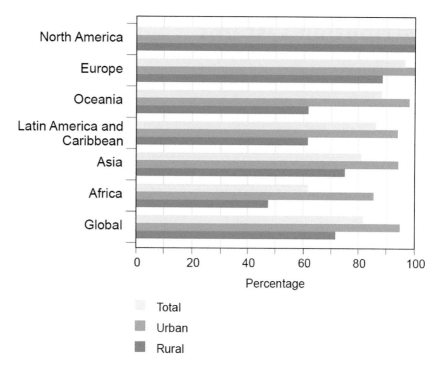

Figure 1.1 Water supply coverage in the world in 2000
Source: WHO/UNICEF/WSSCC, 2000.

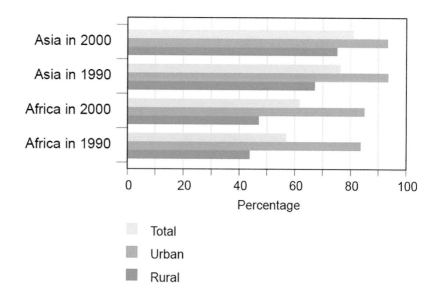

Figure 1.2 Growth of water supply coverage in Africa and Asia between 1990 and 2000
Source: WHO/UNICEF/WSSCC, 2000.

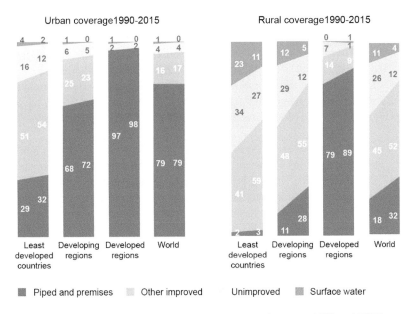

Figure 1.3 Trends in drinking water coverage percentages between 1990 and 2015

Source: WWAP, 2015.

(WWAP, 2015). Expressed in numbers, the figure of approximately one billion people in the world lacking improved access to safe drinking water in the year 2000 was reduced to 663 million in 2015, but still nearly half of these live in Sub-Saharan Africa.

The following are some examples of different water supply standards worldwide:

1) A study carried out in the Netherlands in the late eighties indicates a remarkably low average frequency of water supply interruption; even this long time ago, the chance that no water would run after turning on the tap was once in 14 years (Baggelaar *et al.*, 1988)! Not much has changed in the meantime except that the water consumption has been reduced. Despite the continuously improved level of service, plentiful supply and affordable tariffs, the average domestic water consumption in the Netherlands rarely exceeds 120 litres per capita per day (l/c/d), according to the Dutch association of drinking water companies (VEWIN, 2015).

2) Abu-Madi and Trifunović (2013) report the water consumption in the West Bank governorates of the Palestinian Authority (PA) as between 43 l/c/d in Jenin (274,000 inhabitants), and 162 l/c/d in Jericho (45,000 inhabitants). The PA covers a predominantly water-scarce area and the consumers are therefore well aware that their taps may go dry if kept on longer than necessary. Due to the chronic shortage in supply, the water has to be collected in individual tanks stored on the roofs of houses. Nevertheless, at least some consumers can still afford a consumption rate higher than the one common in the Netherlands.

3)   In their Twelfth Five Year Plan - 2012/2017, the intention of the Government of India was to increase the minimum water supply coverage of 40 l/c/d for the rural population from 35 % to 50 %. This access should be guaranteed '*within their household premises or within 100 metre radius (and within 10 metre elevation in hilly areas) from their households without barriers of social or financial discrimination.*' In addition, the individual household connections in rural areas should grow from 13 % to minimum 35 %, by 2017 (Planning Commission, Government of India, 2013).

All three examples, registered in different parts of the world, reflect three different realities: urban in continental Europe with a direct supply, urban in an arid area of the Middle East with an intermittent supply but more or less continuous water use, and rural in Asia where the water often has to be collected from a distance. Clearly, the differences in the type of supply, water availability at source and overall level of infrastructure all have significant implications for the quantities of water used. Finally, at the extreme end of the scale there is little concern about the frequency of water supply interruptions; the water is fetched in buckets and average quantities are a few litres per head per day, which can be better described as 'a few litres *on* head per day', as Figure 1.4 shows.

The relevance of a reliable water supply system is obvious. The common belief that the treatment of water is the most expensive process in these systems is disproved by many examples. In the case of the Netherlands, the total value of assets of water supply works, assessed in 1988 at a level of approximately EUR 4 billion, shows a proportion where more than a half of the total cost can be attributed to water transport and distribution facilities including service connections, and less than half is apportioned to the raw water extraction and treatment; this is shown in Figure 1.5 (VEWIN, 1990). More recent data given in Figure 1.6 on annual investments, averaging approximately EUR 400 million, in the reconstruction and expansion of these systems consistently show the largest expenditure is on water distribution (VEWIN, 2001; 2015).

The situation in the Netherlands shown in figures 1.5 and 1.6 is not unique and is likely to be found in many other countries, pointing to the conclusion that transport and distribution are dominant processes in any water supply system. Moreover, the data shown include capital investments without exploitation costs, which are costs that can be greatly affected by inadequate design, operation and maintenance of the system, resulting in excessive water and energy losses or deterioration of water quality on its way to consumers. Regarding the first problem, there are numerous examples of water distribution systems in the world where sometimes half of the total production remains unaccounted for, and where a vast quantity of it is physically lost from the system.

Dhaka is the capital of Bangladesh with a population of some 7 million, with 80% of the population being supplied by the local water company and the average consumption is approximately 117 l/c/d (McIntosh, 2003). Nevertheless, less than 5 % of the consumers receive a 24-hour supply, the rest being affected by frequent operational problems. Moreover, water losses are estimated at 40 % of the total production. A simple calculation shows that under normal conditions, with water losses at a level of say 15 %, the same production capacity would be sufficient to

Figure 1.4 The 21st century somewhere in Africa (top) and Asia (bottom)...

supply the entire population of Dhaka with approximately 120 l/c/d, i.e. similar to the average in the Netherlands.

Hence, transport and distribution systems are very expensive even when perfectly designed and managed. Optimisation of design, operation and maintenance has always been, and will remain, the key challenge for any water supply company. Nowadays, this fact is underlined by the population explosion that is

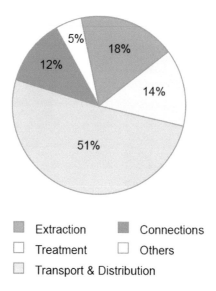

Extraction   Connections

Treatment   Others

Transport & Distribution

Figure 1.5 Structure of assets of the Dutch water supply works in 1988
Source: VEWIN, 1990.

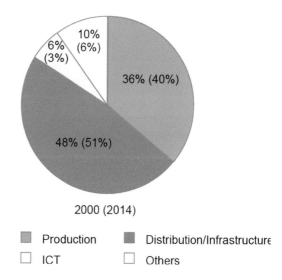

2000 (2014)

Production   Distribution/Infrastructure

ICT   Others

Figure 1.6 Annual investments in the Dutch water supply works in 2000 and 2014
Source: VEWIN, 2001; 2015.

expected to continue in urban areas, particularly in developing and newly indus-
trialised countries in the coming years. According to the prediction shown in
Table 1.1, nearly 6.3 billion people will be living in urban areas of the world by
the year 2050, which is 88% more than in the year 2011, and almost nine times
as many as in 1950. Not surprisingly, the most rapid growth is expected on the

Table 1.1 World population growth 1950–2050

Region	Total population (millions)/urban population (%)				
	1950	*1970*	*2011*	*2030*	*2050*
North America	172/64	231/74	348/82	402/86	447/89
Europe	547/51	656/63	739/79	741/77	719/82
Oceania	13/62	20/71	37/71	47/71	55/73
Latin America and the Caribbean	167/41	286/57	597/79	702/83	751/87
Asia	1403/18	2135/24	4207/45	4868/56	5142/64
Africa	230/14	368/24	1046/40	1562/48	2192/58
Global	2532/29	3696/37	6974/52	8322/60	9306/67

Source: UN, 2012.

Table 1.2 World urban population growth 1970–2025

Areas	Population (millions)/% of total			
	1970	*1990*	*2011*	*2025*
Cities above 10 million inhabitants	39/3	145/6	359/10	630/14
Cities 5 to 10 million inhabitants	109/8	142/6	283/8	402/9
Cities 1 to 5 million inhabitants	244/18	456/20	775/21	1128/24
Cities 0.5 to 1 million inhabitants	128/9	206/9	365/10	516/11
Cities less than 0.5 million inhabitants	833/62	1333/59	1850/51	1967/42
Total	1352/100	2282/100	3632/100	4643/100

Source: UN, 2012.

two most populated continents, Asia and Africa, and in so-called mega-cities, as Table 1.2 shows.

It is not difficult to anticipate the stress on infrastructure that these cities are going to face, with a supply of safe drinking water being one of the major concerns. The goal of an uninterrupted supply has already been achieved in the developed world where the focus has shifted towards environmental issues, but in many less developed countries, this is still a dream.

1.2 Definitions and objectives

1.2.1 Transport and distribution

In general, a water supply system comprises the following processes (Figure 1.7):

1) raw water extraction and transport,
2) water treatment and storage, and
3) clear water transport and distribution.

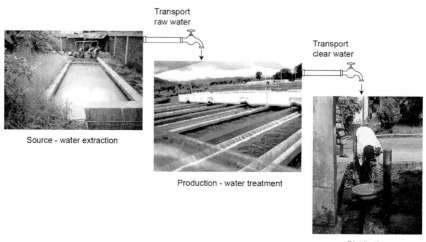

Figure 1.7 Water supply system processes

Transport and distribution are technically the same processes in which the water is conveyed through a network of pipes, stored intermittently and pumped where necessary, in order to meet the demands and pressures in the system; the difference between the two is in their objectives, which influence the choice of system configuration.

Water transport systems comprise main transmission lines with high and fairly constant capacities. Except for potable water, these systems may be constructed for the conveyance of raw or partly treated water. As a part of the drinking water system, these transport lines do not directly serve consumers. They usually connect the clear water reservoir of a treatment plant with some central storage in the distribution area. Interim storage or booster pumping stations may be required in the case of long distances, specific topography or branches in the system.

Branched water transport systems provide water for more than one distribution area forming a regional water supply system. Probably the most remarkable examples of such systems exist in South Korea. The largest of 23 multi-regional and industrial water supply systems supplies 25 million inhabitants of the capital Seoul and its satellite cities. The 860 km-long system of concrete pipes and tunnels in diameters ranging up to 4.3 metres was completed in 2008 to supply an average capacity of 8.3 million cubic metres per day (m^3/d) (K-water, 2015). However, the largest in the world is the famous 'Great Manmade River' transport system in Libya, which is still under construction. Its first two phases were completed in 1994 and 2000, respectively. The approximately 3500 km-long system, which was constructed of concrete pipes of 4 metres in diameter, supplies approximately 3 million m^3/d of water. This is mainly used for irrigation and also partly for water supply for the cities in the coastal area of the country. After all the three remaining phases of construction have been completed, the total capacity provided will be approximately 5.7 million m^3/d. Figure 1.8 gives an impression

Water transport systems

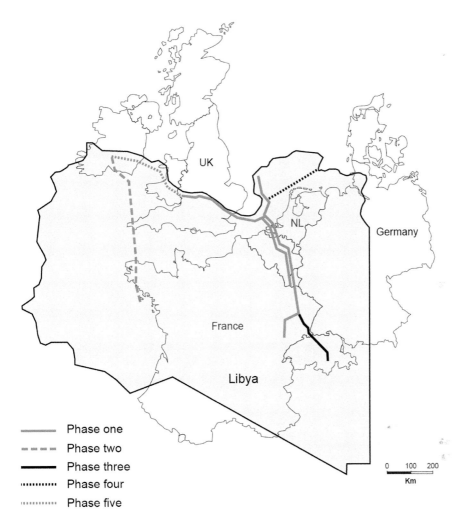

Figure 1.8 The 'Great Manmade River' transport system in Libya

Source: The Management and Implementation Authority of the GMR project, 1989.

of the size of the system by laying the territory of Libya (the grey area) over the map of Western Europe (The Management and Implementation Authority of the GMR project, 1989).

Water distribution systems consist of a network of smaller pipes with numerous connections that supply water directly to the users. The flow variations in such systems are much bigger than in cases of water transport systems. In order to achieve optimal operation, different types of reservoirs, pumping stations, water towers, as well as various appurtenances (valves, hydrants, measuring equipment, etc.) can be installed in the system.

Water distribution systems

The example of a medium-size distribution system in Figure 1.9 shows the looped network of Zanzibar in Tanzania, a town of approximately 230,000 inhabitants. The average supply capacity is approximately 27,000 m³/d (Hemed, 1996). The dotted lines in the figure indicate pipe routes planned for future extensions; this network layout was the basis of a computer model that consisted of some 200

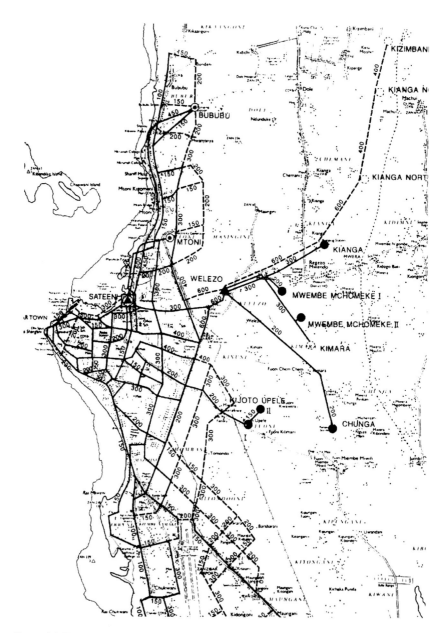

Figure 1.9 The water distribution system in Zanzibar, Tanzania

Source: Hemed, 1996.

pipes and was used effectively in describing the hydraulic performance of the network.

The main objectives of water transport and distribution systems are common:

- the supply of adequate water quantities,
- maintaining the water quality achieved by the water treatment process.

Both of these objectives should be satisfied for all consumers at all times and, bearing in mind the massive scale of such systems, at an acceptable cost. This presumes a capacity of water supply for basic domestic purposes, commercial, industrial and other types of use and, where possible and economically justified, for fire protection.

Speaking in hydraulic terms, sufficient quantity and quality of water can be maintained by adequate pressure and velocity. Keeping pipes always under pressure drastically reduces the risks of external contamination. In addition, conveying the water at an acceptable velocity helps to reduce the retention times, which prevents the deterioration in quality resulting from low chlorine residuals, the appearance of sediments, the growth of microorganisms, etc. Hence, *potable water in transport and distribution systems must always be kept under a certain minimum pressure and for hygienic reasons should not be left stagnant in pipes.*

Considering the engineering aspects, the quantity and quality requirements are met by making proper choices in the selection of components and materials. The system components used for water transport and distribution should be constructed i.e. manufactured from durable materials that are resistant to mechanical and chemical attacks, and at the same time not harmful for human health. Just as importantly, their dimensions should comply with established standards.

Finally, in satisfying the quantity and quality objectives special attention should be paid to the level of workmanship during the construction phase as well as later on, when carrying out the system operation and maintenance. Lack of consistency in any of these indicated steps may result in the pump malfunctioning, leakages, bursts, etc. with the possible consequence of contaminated water.

1.2.2 Piping

Piping is a part of transport and distribution systems that demands major investments. The main components comprise pipes, joints, fittings, valves and service connections. According to the purpose they serve, the pipes can be classified as follows:

The trunk main is the pipe used for the transport of potable water from the treatment plant to the distribution area. Depending on the maximum capacity i.e. demand of the distribution area, the common range of pipe sizes is very wide; trunk mains can have diameters of between a few 100 millimetres and a few meters, in extreme cases. Some branching of the pipes is possible but consumer connections are rare. *Trunk main*

Secondary mains are pipes that form the basic skeleton of the distribution system. This skeleton normally links the main components, sources, reservoirs and pumping stations, and should enable the smooth distribution of bulk flows towards the areas of higher demand. When properly developed, it also supports *Secondary mains* the system operation under irregular conditions (fire events, a major pipe burst or maintenance, etc.). A number of service connections can be provided from these pipes, especially for large consumers. Depending on the size of the network, typical diameters are 150-400 mm.

Distribution mains *Distribution mains* convey water from the secondary mains towards various consumers. These pipes are laid alongside roads and streets with service connections and valves connected to guarantee the required level of supply. In principle, common diameters are between 50-200 mm.

The schematic layout of a distribution network supplying some 350,000 consumers is given in Figure 1.10. The sketch shows the end of the trunk main that connects the reservoir and pumping station with the source well field. The water is pumped from the reservoir through the network of secondary mains of diameters D = 300-600 mm and further distributed by the pipes D = 100 and 200 mm.

Service pipes From the distribution mains, numerous *service pipes* bring the water directly to the consumers. In the case of domestic supplies, the service pipes are generally around 25 mm (1 inch) but other consumers may require a larger size.

The end of the service pipe is the end point of the distribution system. From that point on, two options are possible:

Public connection *Public connection*; the service pipe terminates in one or more outlets and the water is consumed directly. This can be any type of public tap, fountain, etc.

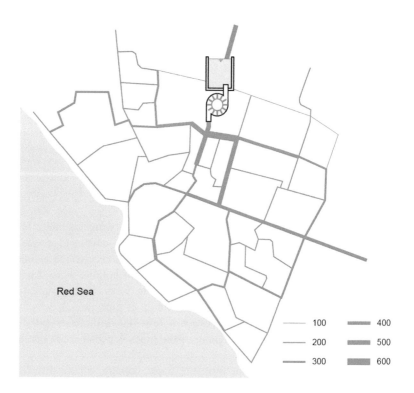

Red Sea

100		400	
200		500	
300		600	

Figure 1.10 The distribution system in Hodaidah, Yemen
Source: Trifunović and Blokland, 1993.

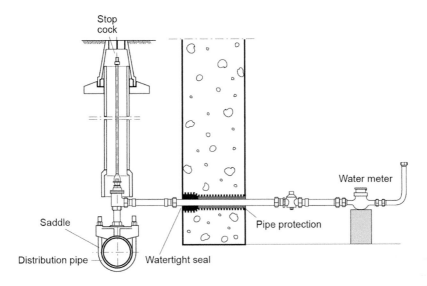

Figure 1.11 Schematic layout of a service connection

Private connection; the service pipe terminates at a stopcock of a private instal- *Private*
lation within a dwelling. This is the point where the responsibility of the water *connection*
supply company usually stops. These can be different types of house or garden
connections, as well as connections for non-domestic use.

One typical domestic service connection is shown in Figure 1.11.

1.2.3 Storage

Clear water storage facilities are a part of any sizable water supply system. They
can be located at source (i.e. the treatment plant), at the end of the transport system or at any other beneficial location in the distribution system, usually at higher
elevations. Reservoirs (or tanks)[1] serve the following general purposes:

- meeting variable supply to the network with constant water production,
- meeting variable demand in the network with the constant supply,
- providing a supply in emergency situations, and
- maintaining stable pressure (if sufficiently elevated).

Except for very small systems, the costs of constructing and operating water storage facilities are comparable to the savings achieved in building and operating
other parts of the distribution system. Without the use of a storage reservoir at the
end of the transport system, the flow in the trunk main would have to match the

---

1 The distinction between the terms 'reservoir' and 'tank' is not always clear and is usually made by
   the storage volume and location. Reservoirs are commonly of larger volume and located at supply
   points/treatment plants, while tanks are typically positioned within distribution networks.

demand in the distribution area at any moment in time, resulting in higher design flows i.e. larger pipe diameters. When operating in conjunction with a storage reservoir, this pipe only needs to be sufficient to convey the average flow, while the maximum peak flow will be supplied by drawing the additional requirement from the balancing volume in the reservoir.

The selection of an optimal site for a reservoir depends upon the type of supply scheme, topographical conditions, the pressure situation in the system, economic aspects, climatic conditions, security, etc. The required volume to meet the demand variations will depend on the daily demand pattern and the way the pumps are operated. Stable consumption over 24 hours normally results in smaller volume requirements than in cases where there is a big range between the minimum and maximum hourly demand. Finally, a proper assessment of needs for supply under irregular conditions can be a crucial decision factor.

Total storage volume in one distribution area commonly covers between 20-50 % of the maximum daily consumption within any particular design year. With additional safety requirements, this percentage can be even higher. See Section 4.2.3 for a further discussion of the design principles. Figure 1.12 shows the total reservoir volumes in a number of cities around the world (Kujundžić, 1996).

The reservoirs can be constructed either:

- underground,
- at ground level, or
- elevated (water towers).

Underground reservoirs (Figure 1.13) are usually constructed in areas where safety or aesthetical issues are important. In tropical or cold climates, preserving the water temperature i.e. water quality could also be considered when choosing such a construction.

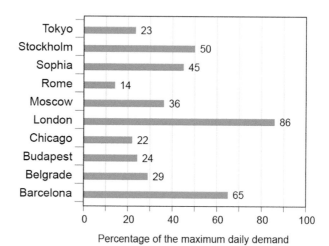

Figure 1.12 Total storage volume in a number of cities around the world
Source: Kujundžić, 1996.

Figure 1.13 An underground reservoir in Lisbon, Portugal

Figure 1.14 Ground level concrete reservoirs in Galápagos, Ecuador

Compared to underground reservoirs, ground-level reservoirs (Figure 1.14) are generally cheaper and offer easier accessibility for maintenance. Both of these types have the same objectives: balancing demand and providing a buffer reserve.

Elevated tanks, also called *water towers*, are typical for predominantly flat ter- *Water towers* rains in cases where the required pressure levels could not have been reached by

Figure 1.15 Water towers in Amsterdam (still in use) and Delft (no longer in use)

positioning the ground tank at some higher altitude. These tanks rarely serve as a buffer in irregular situations; large elevated volumes are generally unacceptable due to economic reasons. The role of elevated tanks is different to that of ordinary balancing or storage reservoirs. The volume here is primarily used for balancing of smaller and shorter demand variations and not for daily accumulation. Therefore, the water towers are often combined with pumping stations, preventing too frequent switching of the pumps and stabilising the pressure in the distribution area at the same time. Two examples of water towers in the Netherlands are shown in Figure 1.15.

In some cases, tanks can be installed at the consumer's premises if:

- the consumers would otherwise cause large fluctuations in water demand,
- the fire hazard is too high,
- back-flow contamination (by the user) of the distribution system has to be prevented, or
- an intermittent water supply is unavoidable.

In cases of restricted supply, individual storage facilities are inevitable. Very often, the construction of such facilities is out of proper control and the risk of contamination is relatively high. Nevertheless, in the absence of other viable alternatives, these are widely applied in arid areas of the world, such as in the Middle East, Southeast Asia or South America. One example of a multi-storage building supplied from a cluster of roof tanks is shown in Figure 1.16 (Trifunović and Abu-Madi, 1999).

A typical individual supply scheme from Sana'a, the Republic of Yemen, in Figure 1.17 shows a ground-level tank with a volume of 1-2 m³, connected to the distribution network. This reservoir receives the water in periods when the pressure in the distribution system is sufficient. The pressure in the house installation is maintained from the roof tank that is filled by a small pump. Both reservoirs

Figure 1.16 Roof tanks in Ramallah, Palestinian Authority
Source: Trifunović and Abu-Madi, 1999.

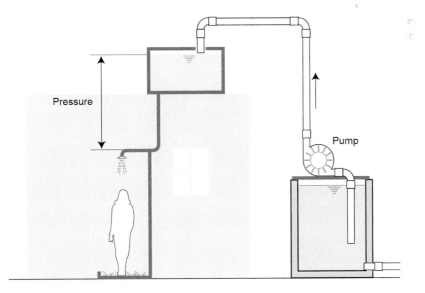

Figure 1.17 Individual storage in water-scarce areas
Source: Trifunović, 1994.

have float valves installed in order to prevent overflow. In more advanced applications, the pump may operate automatically depending on the water level in both tanks. In areas of the town with higher pressure, the roof tanks will be directly connected to the network.

In theory, this kind of supply allows for lower investment in distribution pipes as the individual balancing of demand reduces the peak flows in the system. In addition, generally lower pressures associated with the supply from the roof tanks have a positive effect on leakages. In practice however, the roof tanks are more often a consequence of a scarce resource and/or a poor service level rather than a water demand management tool.

In Europe, roof tanks can be seen in arid areas of the Mediterranean belt. Furthermore, they are traditionally built in homes in the UK. The practice there dates from the 19th century when water supplied to homes from the municipal water companies was intermittent, which is the same reason as in many developing countries nowadays. Such tanks, usually of a few 100 litres, are typically installed under the roof of a family house and are carefully protected from external pollution. Their present role is now less for emergencies and more as small balancing tanks. Furthermore, the roof tanks in the developed world are frequently encountered in large multi-storey buildings, for provision of pressure and for firefighting on the higher floors.

1.2.4   Pumping

Pumps add energy to water. Very often, the pumping operation is closely related to the functioning of the balancing reservoirs. Highly-elevated reservoirs will usually be located at the pressure (i.e. downstream) side of the pumping station in order to be refilled during the periods of low demand. Low-level reservoirs, on the other hand, will be positioned at the suction (i.e. upstream) side of the pumping station that provides supply to the consumers located at higher elevations. Pumps can also be located at the source (Figure 1.18) or anywhere else in the network where additional pressure is required; due to this primary role they are commonly called *booster stations* (Figure 1.19).

Centrifugal flow pumps

Centrifugal flow pumps are mostly used in water distribution. They can be installed in a horizontal (Figure 1.20) or vertical set-up if the available space restricted (see Figure 1.21). The main advantages of centrifugal pumps are low maintenance costs, high reliability, a long lifetime, and simple construction, which all ensure that the water pumped is hygienically pure.

The pump unit is commonly driven by an electrical motor or a diesel engine, the latter being an alternative in case of electricity failures or in remote areas not connected at all to the electricity network. Two groups of pumps can be distinguished with respect to the motor operation:

1)   fixed speed pumps, and
2)   variable speed pumps.

Frequency converter

In the first case, the pump is driven by a motor with a fixed number of revolutions. In the second case, an additional installed device, called the *frequency converter*, controls the impeller rotation thus enabling a more flexible pump operation.

Figure 1.18 Deep well pump at the groundwater source

Figure 1.19 Booster stations

Figure 1.20 Horizontal centrifugal pumps

Figure 1.21 Vertical centrifugal pumps with frequency controllers for variable speed

Variable speed pumps can achieve the same hydraulic effect as fixed speed pumps in combination with a water tower, rendering water towers unnecessary. By changing the speed, these pumps are able to follow the demand pattern within certain limits whilst at the same time keeping almost constant pressure. Consequently, the same range of flows can be covered with a smaller number of units. However, this technology has some restrictions; it cannot cover a large demand variation. Moreover, it involves rather sophisticated and expensive equipment, which is probably the reason why it is predominantly applied in developed countries. With obvious cost-saving effects, variable speed pumps are widely used in the Netherlands where the vast majority of over 200 water towers built throughout the 19th and 20th centuries has been disconnected from operation in the last three decades, being considered uneconomical.

Variable speed pumps

Proper selection of the type and number of pump units is of crucial importance for the design of pumping stations. Connecting pumps in a *parallel arrangement* enables a wider range of flows to be covered by the pumping schedule while with pumps connected in a *serial arrangement* the water can be brought to extremely high elevations. A good choice in both cases guarantees that excessive pumping heads will be minimised, pumping efficiency increased, energy consumption reduced, working hours of the pumps better distributed, and their lifetime extended.

1.3   Types of distribution schemes

With respect to the way the water is supplied, the following distribution schemes can be distinguished:

1) gravity,
2) direct pumping, and
3) combined.

The proper choice is closely linked to the existing topographical conditions.

Gravity schemes make use of the existing topography. The source is, in this case, located at a higher elevation than the distribution area itself. The water distribution can take place without pumping and nevertheless under acceptable pressure. The advantages of this scheme are:

Gravity schemes

- no energy costs,
- simple operation (fewer mechanical components, no power supply needed),
- low maintenance costs,
- slower pressure changes, and
- a buffer capacity for irregular situations.

However, as much as they can help in creating pressure in the system, the topographical conditions may also obstruct future supplies. Due to the fixed pressure range, the gravity systems are less flexible with regard to extensions. Moreover, they require larger pipe diameters in order to minimise pressure losses. The main operational concern is capacity reduction that can be caused by air entrainment.

Direct pumping schemes

In *direct pumping schemes*, the system operates without storage provision for demand balancing. The entire demand is directly pumped into the network. As the pumping schedule has to follow variations in water demand, the proper selection of units is important in order to optimise the energy consumption. Reserve pumping capacity for irregular situations should also be planned.

Advantages of the direct pumping scheme are opposite to those of the gravity scheme. With good design and operation, any pressure in the system can be reached. However, these are systems with a rather complicated operation and maintenance and they are dependent on a reliable power supply. Additional precautions are therefore necessary, such as an alternative source of power supply, automatic mode of pump operation, stock of spare parts, etc.

Combined schemes

Combined schemes assume an operation with pumping stations and demand balancing reservoirs. Part of the distribution area may be supplied by the direct pumping and the other part by gravity. A considerable storage volume is needed in this case but the pumping capacities will be below those in the direct pumping scheme. These combined systems are common in hilly distribution areas.

Pressure zones

The prevailing topography can lead to the use of *pressure zones*. By establishing different pressure zones, savings can be obtained in supplying water to the various elevations at lower pumping costs and in the use of lower-class piping, due to the lower pressure. Technically, the pressure zones may be advantageous in preventing too high pressures in lower parts of the network (pressure-reducing valves may be used), or providing sufficient pressures in the higher parts (by pumping) when the source of supply is located in the lower zone.

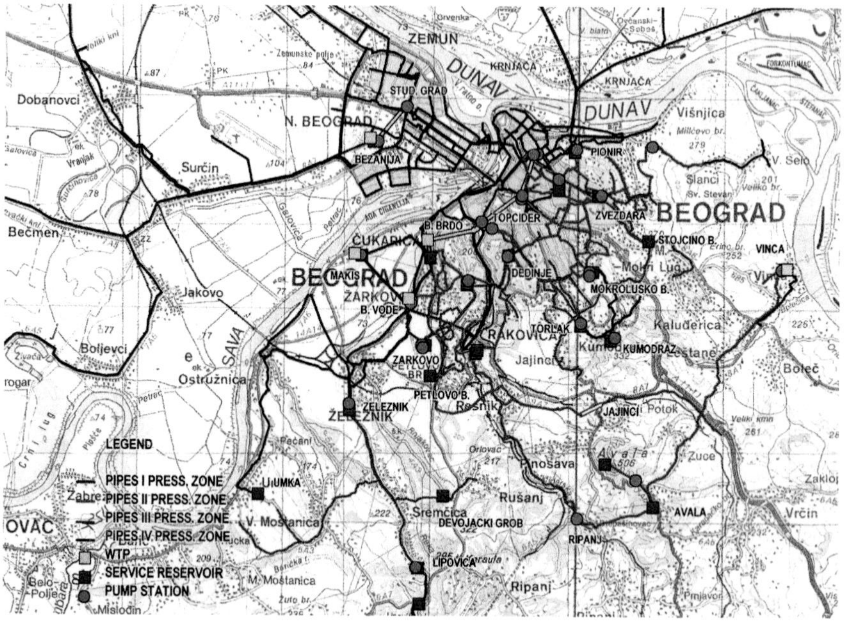

Figure 1.22 The water supply network in Belgrade, Serbia
Source: BWS, 2003.

An interesting application of zoning can be seen in Stuttgart, Germany. The distribution network for approximately 600,000 consumers is located in a terrain with an elevation difference of some 300 m between the lowest and the highest points. It is divided into nine pressure zones and further split into 64 smaller and individually monitored distribution areas, so-called pressure management areas (PMAs). In each of these sub-areas the pressure range is kept between 20-60 metres water column (mwc). The main advantage of such a system is that a major failure is virtually impossible. In the case of a calamity in any of the sub-areas, an alternative supply from the neighbouring areas can be arranged by adjusting the network operation. However, a centralised and very carefully synchronised control of the system that consists of a total of 2500 km of network pipes, 39 pumping stations, 44 elevated tanks and 88 reservoirs is essential to achieve this level of operation (Netze BW, 2010).

In another, more conventional, application of pressure zoning, the water supply system in Belgrade in Serbia, shown in Figure 1.22, supplies the population of approximately 1.4 million with approximately one million m³/day. The system comprises four zones supplied from five water treatment plants and operated by 36 pumping stations and 27 service reservoirs that have a total volume of 240,000 m³ (BWS, 2003).

1.4   Network configurations

Depending on the way the pipes are interconnected, the following network configurations can be distinguished:

1) serial,
2) branched,
3) looped (or grid), and
4) combined.

A serial network is a network without branches or loops, the simplest configuration of all (Figure 1.23 A). It has one source, one end and a couple of intermediate nodes (demand points). Each intermediate node connects two pipes: supply i.e. an upstream pipe and distribution i.e. a downstream pipe. The flow direction is fixed from the source to the end of the system. *Serial networks*

These networks characterise very small (rural) distribution areas and although rather inexpensive, they are not common due to extremely low reliability and quality problems caused by water stagnation at the end of the system. When this configuration is used for water transport, the large diameters and lengths of the pipes will cause a drastic increase in the construction costs. Where reliability of supply is of greater concern than the construction cost, parallel lines will be laid.

A branched network is a combination of serial networks. It usually consists of one supply point and several ends (Figure 1.23 B). In this case, the intermediate nodes in the system connect one upstream pipe with one or several downstream pipes. A fixed flow direction is consequently generated by the distribution from the source to the ends of the system. *Branched networks*

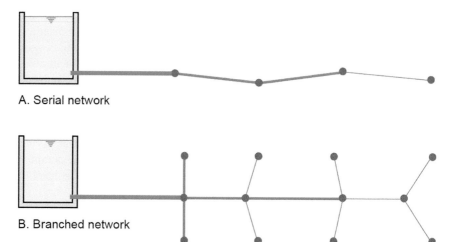

A. Serial network

B. Branched network

Figure 1.23 Serial and branched network configurations

Branched networks are adequate for small communities, bearing in mind acceptable investment costs. However, the main disadvantages remain:

- low reliability,
- potential danger of contamination caused by large parts of the network being without water during irregular situations,
- accumulation of sediments due to stagnation of the water at the system ends ('dead' ends), occasionally resulting in taste and odour problems, and
- a fluctuating water demand producing rather high pressure oscillations.

Looped networks   *Looped (or grid) networks* (Figure 1.24 A) consist of nodes that can receive water from more than one side. This is a consequence of the looped structure of the network formed in order to eliminate the disadvantages of branched systems. A looped layout can be developed from a branched system by connecting its ends either at a later stage, or initially as a set of loops. The problems encountered in branched systems will be eliminated in the following circumstances:

- the water in the system flows in more than one direction and a long lasting stagnation no longer easily occurs,
- during the system maintenance, the area concerned will continue to be supplied by water flowing from other directions; in the case of pumped systems, a pressure increase caused by a restricted supply may even promote this, and
- water demand fluctuations will produce less effect on pressure fluctuations.

Looped networks are hydraulically far more complicated than serial or branched networks. The flow pattern in such a system is predetermined not only by its layout but also by the system operation. This means that the location of critical

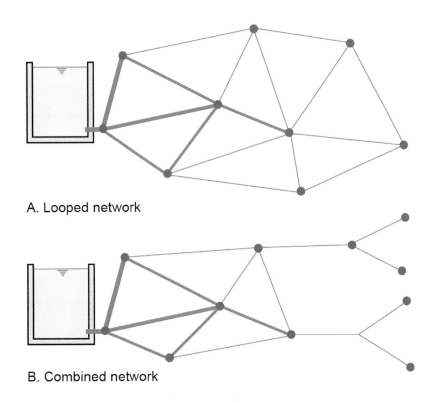

A. Looped network

B. Combined network

Figure 1.24 Looped and combined network configurations

pressures may vary over time. In the case of supply from more than one source, the analysis becomes even more complex.

Looped networks are more expensive both in investment and costs of operation. They are appropriate solutions predominantly for those (urban or industrial) distribution areas that require a high reliability of supply.

A combined network is the most common type of network in large urban areas. A looped structure forms the central part of the system while the supply on the outskirts of the area is provided through a number of extended lines (Figure 1.24 B).

Combined networks

All the advantages and disadvantages of combined systems relating to both branched and looped systems have already been discussed.

Chapter 2

Water demand

2.1 Terminology

Water conveyance in a water supply system depends on the rates of production, delivery, consumption and leakage (Figure 2.1).

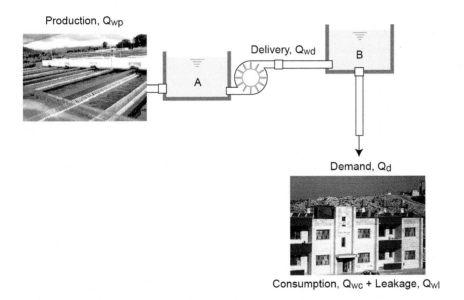

Figure 2.1 Flows in a water supply system

Water production (Q_{wp}) takes place at water treatment facilities. It normally has a constant rate that depends on the purification capacity of the treatment installation. The treated water ends up in a clear water reservoir (reservoir A in Figure 2.1, constructed as a buffer for possible interruptions in the purification process) from where it is supplied to the system. *Water production*

Water delivery (Q_{wd}) starts from the clear water reservoir of the treatment plant. Supplied directly to the distribution network, the generated flow will match the demand patterns. When the distribution area is located far away from the treatment plant, the water is likely to be transported to another reservoir which has the primary role of balancing the demand variations (reservoir B in Figure 2.1); this reservoir is *Water delivery* usually constructed at the entry point to the distribution network. In principle, this delivery is made at the same constant flow rate as the water production.

Water consumption

Water consumption (Q_{wc}) is the quantity directly utilised by consumers. This generates variable flows in the distribution network caused by many factors: users' needs, climate, source capacity, etc.

Water leakage

Water leakage (Q_{wl}) is the amount of water physically lost from the system. The generated flow rate is in this case more or less constant and depends on overall conditions in the system.

Water demand

In theory, the term *water demand* (Q_d) coincides with water consumption. In practice however, the demand is often monitored at supply points where the measurements include water leakage, as well as the quantities used to refill the balancing tanks that may exist in the system. In order to avoid false conclusions, a clear distinction between the measurements at various points of the system should always be made. It is commonly agreed that $Q_d = Q_{wc} + Q_{wl}$. Furthermore, when supply is calculated without having an interim water storage, i.e. water goes directly to the distribution network, $Q_{wd} = Q_d$, otherwise: $Q_{wd} = Q_{wp}$.

Water demand is commonly expressed in cubic meters per hour (m³/h) or per second (m³/s), litres per second (l/s), or mega litres per day (Ml/d). Typical Imperial units are cubic feet per second (ft³/s), gallon per minute (gpm) or mega gallon per day (mgd)[1]. The mean value derived from annual demand records represents

Average demand
Specific demand

the *average demand*. Divided by the number of consumers, the average demand becomes the *specific demand* (or *unit consumption per capita*). In this case the common units are litres per capita per day (l/c/d or lpcpd).

Apart from ignoring losses of water and revenues, the demand figures can often be misinterpreted due to lack of information regarding the consumption by various categories. Table 2.1 shows the difference in the level of specific demand depending on what is, or is not, included in the figure.

Table 2.1 Water demand in the Netherlands in 2014

	Annual (10⁶ m³)	Q_d (l/c/d)*
Total delivery by water companies (excl. production losses)	1310	213
Drinking water delivered for distribution	1173	191
Drinking water with collected revenues	1108	180
Households + business market	1068	174
Households alone	783	127

* Divided by total population of 16.83 million (Source: CBS, 2015).

Source: VEWIN, 2015.

Accurate forecasting of water demand is crucial when analysing the hydraulic performance of a water distribution system. Numerous factors affecting the demand are determined from the answers to these three basic questions:

1. *For which purpose is the water used?* The demand is affected by a number of consumption categories: domestic, industrial, tourism, etc.

---

1 A general unit conversion table is given in Appendix 9. See also Spreadsheet Lesson 8-1 (Appendix 7).

2. *Who is the user?* Water use within the same category may vary due to differ-
    ent cultures, education, age, climate, religion, or technological process.
3. *How valuable is the water?* The water may be used under circumstances that
    restrict the demand: scarce source (quantity/quality), poor access (no direct
    connection, fetching from a distance), low income of consumers, etc.

Answers to the above questions reflect on the quantities and moments in time
when the water will be used, resulting in a variety of demand patterns. Analysing
or predicting these patterns is not always an easy task. When the field information
is lacking, uncritical adoption of other experiences is the wrong approach; each
case is independent and the conclusions drawn are only valid for local conditions.

Variations in water demand are particularly visible in developing countries
where prosperity is predominantly concentrated in a few major, usually over-
crowded, cities with peripheral areas often having restricted access to drinking
water. These parts of the system will be supplied from public standpipes, individual wells or tankers, which causes substantial differences in consumption levels
within the same distribution area. Figure 2.2 shows average specific consumption
for a number of large cities in Asia.

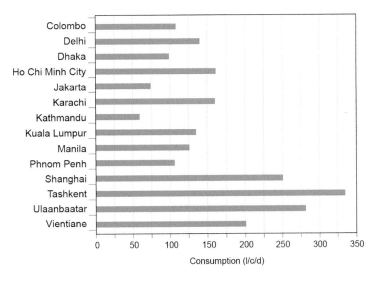

Figure 2.2 Specific consumption in a number of Asian cities

Source: McIntosh, 2003.

Comparative figures for consumption in smaller urban areas are generally lower,
resulting from a range of problems that cause intermittent supply, namely long
distances, electricity failures, pipe bursts, polluted groundwater in deep wells, etc.

A water demand survey was conducted for the region around Lake Victoria,
covering parts of Uganda, Tanzania and Kenya. The demand where there is a piped
supply (the water is tapped at home) was compared with the demand in unpiped
systems (no house connection is available). The results are shown in Table 2.2.

Table 2.2 Specific demand around Lake Victoria in Africa

	Piped (l/c/d)	Unpiped (l/c/d)
Average for the entire region	45	22
Average for urban areas (small towns)	65	26
Average for rural areas	59	8
Part of the region in Uganda	44	19
Part of the region in Tanzania	60	24
Part of the region in Kenya	57	21

Source: IIED, 2000

Non-revenue water

An unavoidable component of water demand is *non-revenue water* (NRW), the water that is supplied 'free of charge'. In a considerable number of transport and distribution systems in developing countries this is the most significant 'consumer' of water, accounting sometimes for over 50 % of the total water delivery.

Causes of NRW differ from case to case. Most often it is a water leakage caused by inadequate maintenance of the network. Other non-physical losses are related to the water that is supplied and has reached the taps, but is not registered or paid for (under-reading of water meters, illegal connections, washing streets, flushing pipes, etc.). This topic is discussed in details in Section 6.1.3.

2.2   Consumption categories

2.2.1   Water use by various sectors

Water consumption is initially split into domestic and non-domestic components. The bulk of non-domestic consumption relates to the water used for agriculture, occasionally delivered from drinking water supply systems, and for industry and other commercial uses (shops, offices, schools, hospitals, etc.). The ratio between the domestic and non-domestic consumption in Europe is shown in Figure 2.3. The figure reflects quite different ratios within a wide range of specific demands influenced by various population size, level and type of industry, climate, etc.

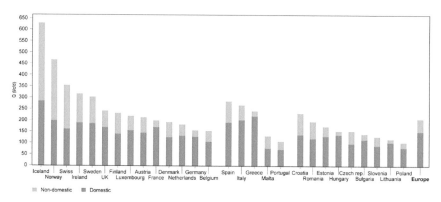

Figure 2.3 Domestic and non-domestic consumption in Europe

Source: EUREAU, 2009.

In the majority of developing countries, agricultural and domestic water consumption is predominant compared to commercial water use, as the example in Table 2.3 shows. However, due to a poor coverage and general shortages the water for agriculture is rarely supplied from a drinking water system.

In warm climates, the water used for irrigation is generally the major component of total consumption; Figure 2.4 shows an example of a number of European countries. Those in the South, such as Greece, Portugal, Spain, and Italy, use a high percentage of their water resources for irrigation. On the other hand, many highly industrialised countries in continental Europe use large quantities of water, often of drinking quality, for cooling in electricity production; typical examples are France, Germany, and Belgium, which all use more than 50% of their water resources for this purpose. Austria and UK also have relatively high water demand for cooling, yet this is not the case in Norway and Denmark. Striving for more efficient irrigation methods, industrial processes using alternative sources and recycling water have been and still are a concern in developed countries for the last few decades; the domestic water use in many situations accounts for just a small proportion of overall water resources countrywide.

Table 2.3 Domestic vs. non-domestic consumption in a number of African states

Country	Agriculture (%)	Industry (%)	Domestic (%)
Angola	76	10	14
Botswana	48	20	32
Lesotho	56	22	22
Malawi	87	3	10
Mozambique	89	2	9
South Africa	62	21	17
Zambia	77	7	16
Zimbabwe	79	7	14

Source: SADC, 1999.

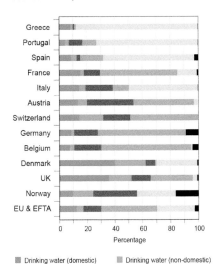

Figure 2.4 Water resources use in Europe

Source: EUREAU, 2009.

2.2.2   *Domestic consumption*

Domestic water consumption is intended for toilet flushing, bathing and shower-ing, laundry, dishwashing and for other less water-intensive or less frequent pur-poses: cooking, drinking, gardening, car washing, etc. The example in Figure 2.5 shows rather wide variation in the specific domestic consumption in some indus-trialised countries. Nevertheless, in all the cases indicated, 50-80 % of the total consumption appears to be utilised in bathrooms and toilets.

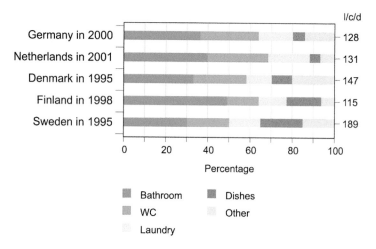

Figure 2.5 Structure of domestic water use in Europe

Sources: EEA; BGW; VEWIN.

A more detailed structure of domestic consumption in the Netherlands in 2013 at an average specific value of 119 l/c/d is shown in Figure 2.6 with the same conclusion: a minimum of 77 % of the total specific consumption takes place in the bathroom.

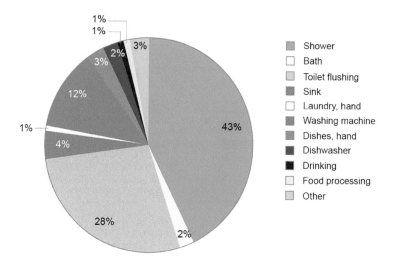

Figure 2.6 Structure of domestic water use in the Netherlands in 2013

Source: VEWIN, 2014.

The habits of different population groups with respect to water use were further studied by comparing the following four factors: number of users per house connection, different age categories, socio-economic class, and major regions in the country (Van Thiel, 2014). These results are shown in Figure 2.7.

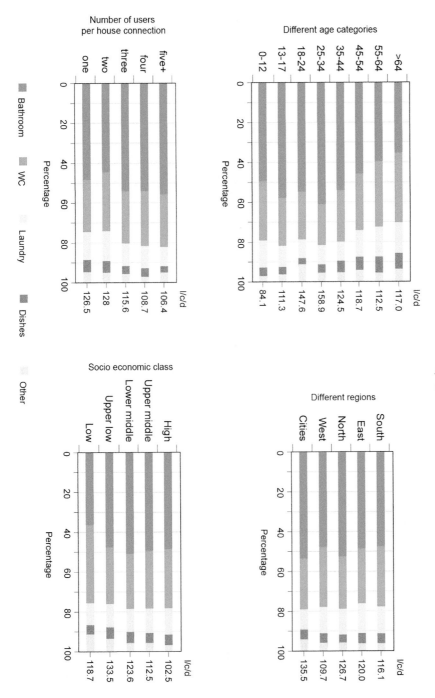

Figure 2.7 Structure of domestic consumption in the Netherlands

Source: Van Thiel, 2014.

The bar charts prove that even with the detailed statistics available, conclusions about global trends may be difficult. In general, the consumption outside Dutch cities is lower, which may have a linkage with the specific lifestyle in urban areas. Nonetheless, interesting findings from the graphs include seemingly more frequent toilet use and less frequent shower use exercised by older groups, larger families with a lower consumption per capita, and a somewhat surprising reduction in specific demand with the increase in socio-economic class, etc.

The trends in domestic water use in the Netherlands in the last few decades clearly point to the reduced specific consumption, which influences the overall demand growth negatively despite the growth in population and the number of households. This is illustrated in Figure 2.8 that shows the specific consumption reduction of some 12 % in the period 1990-2013, even although the population increased at a similar rate, and the number of households increased by nearly 25% (VEWIN, 2015).

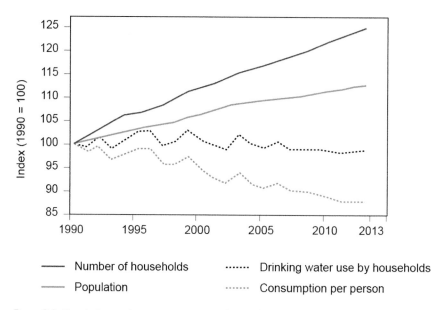

Figure 2.8 Population and consumption growth trends in the Netherlands
Source: VEWIN, 2015.

In cases where there is an individual connection to the system, the structure of domestic consumption in water-scarce areas may well look similar but the quantity of water used for particular activities will be minimized. Apart from the change in habits, this is also a consequence of low pressures in the system directly affecting the quantities used for showering, gardening, car washing, etc. In addition, the water company may be forced to ration the supply by introducing regular interruptions. In these situations consumers will normally react by constructing individual tanks. In urban areas where supply from individual tanks takes place, the amounts of water available commonly vary between 50-100 l/c/d.

2.2.3 Non-domestic consumption

Non-domestic or commercial water use occurs in industry, agriculture, institutions and offices, tourism, etc. Each of these categories has its specific water requirements.

Industry

Water in industry can be used for various purposes: as a part of the final product, for the maintenance of manufacturing processes (cleaning, flushing, sterilisation, conveying, cooling, etc.) and for the personal needs of employees (usually comparatively marginal). The total quantities will largely depend on the type of industry and technological process. They are commonly expressed in litres per unit of product or raw material. Table 2.4 gives an indication for a number of industries; an extensive overview can be found in HR Wallingford (2003).

Table 2.4 Industrial water consumption

Industry	Litres per unit product
Carbonated soft drinks*	1.5 – 5 per litre
Fruit juice*	3 – 15 per litre
Beer*	4 – 22 per litre
Wine	1 – 4 per litre
Fresh meat (red)	1.5 – 9 per kg
Canned vegetables/fruits	2 – 27 per kg
Bricks	15 – 30 per kg
Cement	4 per kg
Polyethylene	2.5 – 10 per kg
Paper	4** – 35 per kg
Textiles	100 – 300 per kg
Cars	2500 – 8000 per car

* Largely dependent on the packaging and cleaning of bottles
** Recycled paper

Adapted from: HR Wallingford, 2003.

Agriculture

Water consumption in agriculture is mainly determined by irrigation and livestock needs. In peri-urban or developed rural areas, this demand may also be supplied from the local drinking water distribution system.

The amounts required for irrigation purposes depend on the plant species, stage of growth, type of irrigation, soil characteristics, climatic conditions, etc. These quantities can be assessed either from records or by simple measurements. A number of methods are available in literature to calculate the consumption based on meteorological data (Evaporation pan, Blaney-Criddle, Penman/Penman-Monteith, etc.). According to Brouwer and Heibloem (1986), the consumption is unlikely to exceed a monthly mean of 15 mm per day, which is equivalent to 150 m^3/d per hectare. Approximate values per crop are given in Table 2.5 based on optimal precipitation rates.

Table 2.5 Seasonal crop water needs

Crop	Season (days/year)	Consumption (mm*/season)
Bananas	300 – 365	1200 – 2200
Beans	75 – 110	300 – 500
Cabbages	120 – 140	350 – 500
Citrus fruit	240 – 365	900 – 1200
Corn	80 – 180	500 – 800
Potatoes	105 – 145	500 – 700
Rice	90 – 150	450 – 750
Sunflowers	125 – 130	600 – 1000
Tomatoes	135 – 180	400 – 800
Wheat	120 – 150	450 – 650

* 1 mm of precipitation is equivalent to 1 litre of water over a surface of 1 m^2.

Source: Brouwer and Heibloem, 1986.

The volume of water required for livestock depends on the sort and age/weight of the animal, as well as climatic/seasonal conditions. The size of the stock and type of production also play a role. For example, the water consumption for milking cows suggested in the state of Ontario in Canada is 68-155 l/d per animal, whilst dry cows and bulls typically only need between 22 and 54 l/d, as indicated in Table 2.6 (Ward and McKague, 2007). Furthermore, for poultry, the table shows that the quantity of water required grows significantly with the birds' age and the ambient temperature.

Table 2.6 Animal water consumption

Animal	Litres per day
Milk cows (production of 14 – 46 kg milk/day)	68 – 155
Dry cows, bulls	22 – 54
Feeder pigs (weight range 23 – 100 kg)	3.2 – 10
Horses (weight range 228 – 683 kg)	13 – 59
Feeder lambs (weight range 27 – 50 kg)	3.6 – 5.2
Chicken broilers (1 – 4 weeks old, per 1000 birds)	50 – 260 (at 21°C)
	50 – 415 (at 32°C)
Chicken broilers (5 – 8 weeks old, per 1000 birds)	345 – 470 (at 21°C)
	550 – 770 (at 32°C)
Turkeys (1 – 7 weeks old, per 1000 birds)	38 – 327 (10 – 21°C)
	38 – 448 (27 – 35°C)
Turkeys (8 – 14 weeks old, per 1000 birds)	403 – 737 (10 – 21°C)
	508 – 1063 (27 – 35°C)
Turkeys (15 – 21 weeks old, per 1000 birds)	747 – 795 (10 – 21°C)
	1077 – 1139 (27 – 35°C)

Adapted from: Ward and McKague, 2007.

Institutions

Commercial consumption in restaurants, shops, schools and other institutions can be assessed as the total supply divided by the number of consumers (employees, pupils, patients, etc.). Accurate figures should be available from local records at water supply companies. Some indications of unit consumption are given in Table 2.7. These assume an individual connection with indoor water installations and waterborne sanitation, and are only relevant during working days.

Table 2.7 Water consumption in institutions and businesses

Premises	Consumption
Schools	25 – 75 l/d per pupil
Hospitals	350 – 500 l/d per bed
Laundries	8* – 60 litre per kg washing
Small businesses	25 l/d per employee
Retail shops/stores	100 – 135 l/d per employee
Offices	65 l/d per employee

* Recycled water used for rinsing

Adapted from: HR Wallingford, 2003.

Tourism

Tourist and recreational activities may also have a considerable impact on water demand. The quantities per person (or per bed) per day vary enormously depending on the type and category of accommodation; in luxury hotels, for instance, this demand can go well above 100 l/c/d, which would then normally include the water used for swimming pools, gardening of green areas, golf courses, etc. The tourist consumption also has to be assessed from the perspective of average occupancy of the guest accommodation throughout the year. Table 2.8 shows average figures in Benidorm, which is one of the most popular tourist resorts in Spain.

Table 2.8 Tourist water consumption in Benidorm, Spain

Accommodation	Total beds	Annual occupancy (%)	Average consumption (m³/month)	Specific consumption (l/c/d)
Camping sites	–	–	–	84
1-star hotel	903	56	217	174
2-star hotel	6074	86	1874	194
3-star hotel	19,867	88	2435	287
4-star hotel	9281	80	3340	361

Source: Rico-Amoros et al., 2009.

Miscellaneous groups

Water consumption that does not belong to any of the above-listed groups can be classified as miscellaneous. These are the quantities used for firefighting, public purposes (washing streets, maintaining green areas, public fountains, etc.), maintenance of water and sewage systems (cleansing, flushing mains) or other specific uses (military facilities, sports complexes, zoos, etc.). Sufficient information on water consumption in such cases should be available from local records. Sometimes this demand is unpredictable and can only be estimated on an empirical or statistical basis. For example, in the case of firefighting, the water use is not recorded and measurements are difficult because it is not known in advance when and where the water will be needed. Provision for this purpose will be planned with respect to potential risks, which is a matter for discussion between the municipal fire department and water supply company.

On average, these consumers do not contribute substantially to the overall demand. Very often they are neither metered nor accounted for and thus classified as NRW.

Problem 2-1

A water supply company has delivered an annual quantity of 80,000,000 m³ to a city of 1.2 million inhabitants. Find out the specific demand in the distribution area. In addition, calculate the domestic consumption per capita with leakage from the system estimated at 15 % of the total supply, and billed non-domestic consumption of 20,000,000 m³/y.

Answer

Gross specific demand can be determined as:

$$Q_{avg} = \frac{80,000,000 \times 1000}{1,200,000 \times 365} \approx 183 \; l/c/d$$

The leakage of 15 % of the total supply amounts to an annual loss of 12 million m³. Reducing the total figure further for the registered non-domestic consumption yields the annual domestic consumption of 80 − 12 − 20 = 48 million m³, which is equal to a specific domestic consumption of approximately 110 l/c/d.

Self-study: Workshop problems 1.1.1 and 1.1.2 (Appendix 1)
Spreadsheet Lesson 8-1 (Appendix 7)

2.3   Water demand patterns

Each consumption category can be considered not only from the perspective of its average quantities but also with respect to the timetable of when the water is used.

Demand variations are commonly described by the *peak factors*. These are the ratios between the demand at particular moments and the average demand for the observed period (hour, day, week, year, etc.). For example, if the demand registered during a particular hour was 150 m³ and for the whole day (24 hours) the total demand was 3000 m³, then the average hourly demand of 3000 / 24 = 125 m³/h would be used to determine the peak factor for the hour, which would be 150 / 125 = 1.2. Other ways of peak demand representation are either as a percentage of the total demand within a particular period (150 m³ for the above hour is equal to 5 % of the total daily demand of 3000 m³), or simply as the unit volume per hour (150 m³/h).

Human activities have periodic characteristics and this applies to water use. Hence, the average water quantities from the previous paragraph are just indications of total requirements. Equally relevant for the design of water supply systems are consumption peaks that appear during one day, week or year. A combination of these maximum and minimum demands defines the absolute range of flows that are to be delivered by the water company.

Time-wise, we can distinguish the *instantaneous, daily (diurnal), weekly* and *annual (seasonal)* patterns in various areas (home, building, district, town, etc.). The larger the area is, the more diverse the demand pattern will be as it then represents a combination of several consumption categories, including leakage.

2.3.1   Instantaneous demand

Instantaneous demand (in some literature *simultaneous demand*) is caused by a small number of consumers during a short period of time: a few seconds or minutes. Assessing this sort of demand is the starting point in building up the demand pattern of any distribution area. In addition, the instantaneous demand is directly relevant for the network design in small residential areas (tertiary networks and house installations). The demand patterns of such areas are much more unpredictable than the demand patterns generated by a larger number of consumers. *The smaller the number of consumers involved, the less predictable the demand pattern will be.*

The following *hypothetical* example illustrates the relation between the peak demands and the number of consumers.

Instantaneous demand

Table 2.9 Example of domestic unit water consumption

Activity	Total quantity (litres)	Duration (minutes)	Q_{ins} (l/h)
A - Toilet flushing	8	1	480
B - Showering	50	6	500
C - Hand washing	2	1/2	240
D - Face and teeth	3	1	180
E - Laundry	60	6	600
F - Cooking	15	5	180
G - Dishes	40	6	400
H - Drinking	1/4	1/20	300
I - Other	5	2	150

A list of typical domestic water activities with provisional unit quantities utilised during a particular period of time is shown in Table 2.9. Parameter Q_{ins} in the table represents the average flow obtained by dividing the total quantity by the duration of the activity, converted into litres per hour. For example, 'Activity A-Toilet flushing' is in fact refilling of the toilet cistern. In this case there is a volume of 8 l, within say one minute after the toilet has been flushed. In theory, to be able to fulfil this requirement, the pipe that supplies the cistern should allow a flow of $8 \times 60 = 480$ l/h within one minute. This flow is thus needed within a relatively short period of time and is therefore called the *instantaneous flow*.

Instantaneous flow

Although the exact moment of water use is normally unpredictable, it is well known that there are periods of the day when it happens more frequently. For most people this is in the morning after they wake up, in the afternoon when they return from work or school, or in the evening before they go to sleep.

Considering a single housing unit, it is not reasonable to assume a situation in which all water-related activities from the above table are executed simultaneously. For example, in the morning, a combination of activities A, B, D and H might occur at the same moment. If this is the assumed maximum demand during the day, the maximum instantaneous flow equals the sum of the flows for these four activities. Hence, the pipe that provides water for the house has to be sufficiently large to convey the flow of $480 + 500 + 180 + 300 = 1460$ l/h.

Instantaneous peak factor

With an assumed specific consumption of 120 l/c/d and, say, four people living together, the *instantaneous peak factor* will be:

$$pf_{ins} = \frac{1460}{\dfrac{120 \times 4}{24}} = 73$$

Thus, there is at least one brief moment within 24 hours when the instantaneous flow to the house is 73 times higher than the average flow of the day.

Applying the same logic to an apartment building or residential area of several houses, one can assume that all the tenants use the water there in a similar way and at a similar moment, but never in exactly the same way *and* at exactly the same moment. Again, the maximum demand of the building/area occurs in the morning. This could consist of, for example, toilet flushing and showering in say three apartments/houses, hand washing in two, teeth brushing in six, doing the laundry in two and drinking water in one. The maximum instantaneous flow out of such a consumption scenario case would be $3A + 3B + 2C + 6D + 2E + 1H = 6000$ l/h, which is the capacity to be provided by the pipe that supplies the building/area. Assuming the same specific demand of 120 l/c/d and for a total of 40 occupants, the instantaneous peak factor is:

$$pf_{ins} = \frac{6000}{\dfrac{120 \times 40}{24}} = 30$$

Any further increase in the number of consumers will cause a further lowering of the instantaneous peak factor, up to a level where this factor becomes independent of the growth in the number of consumers. As a consequence, some large diameter pipes that have to convey water for possibly 100,000 consumers would probably be designed based on a rather low instantaneous peak factor, which in this example is set at 1.4.

A *simultaneity diagram* can be obtained by plotting the instantaneous peak factors against the corresponding number of consumers. The three points from the above example, interpolated exponentially, will yield the logarithmic graph shown in Figure 2.9.

Simultaneity diagram

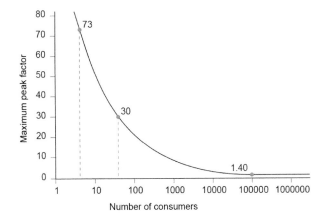

Figure 2.9 Simultaneity diagram (example)

In practice, the simultaneity diagrams are determined from a field study for each particular area. Sometimes, a good approximation is achieved by applying a mathematical formula; the equation: $pf_{ins} \approx 126 \times e^{(-0.9 \times \log N)}$ where N represents the number of consumers, describes the curve in Figure 2.9. Furthermore, the simultaneous curves can be diversified based on various standards of living i.e. type of accommodation, as Figure 2.10 shows.

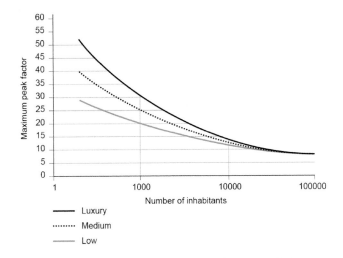

Figure 2.10 Simultaneity diagram for three categories of accommodation

In most cases, the demand patterns of more than a few thousand people are fairly predictable. This eventually leads to the conclusion that the water demand of larger group of consumers will, in principle, be evenly spread over a period of time that is longer than a few seconds or minutes. This is illustrated in the 24-hour demand diagram shown in Figure 2.11 for the northern part of Amsterdam. In this example there were nearly 130,000 consumers, and the measurements were executed at one-minute intervals.

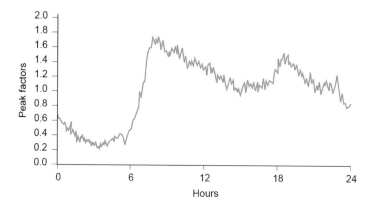

Figure 2.11 Demand pattern in northern Amsterdam

Source: Municipal Water Company Amsterdam, 2002.

A one-hour time step is commonly accepted for practical purposes and the instantaneous peak factors within this period will be represented by a single value called the *hourly (or diurnal) peak factor*, as shown in Figure 2.12.

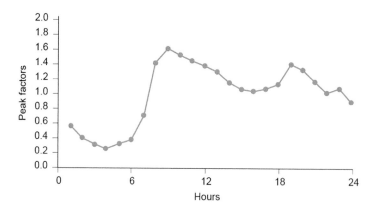

Figure 2.12 Instantaneous demand from Figure 2.11 averaged by the hourly peak factors

There are however extraordinary situations when the instantaneous demand may substantially influence the demand pattern, even in the case of large numbers of consumers.

Figures 2.13 and 2.14 show the demand pattern (in m³/min) during the TV broadcasting of two football matches when the Dutch national team played against Saudi Arabia and Belgium at the 1994 World Cup in the United States of America. The demand was observed in a distribution area of approximately 135,000 people.

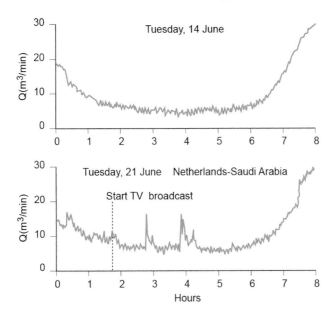

Figure 2.13 Normal night-time demand and demand during a football match
Source: Water Company 'N-W Brabant', NL, 1994.

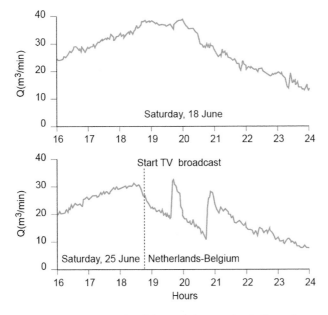

Figure 2.14 Normal evening demand and demand during a football match
Source: Water Company 'N-W Brabant', NL, 1994.

The excitement of the viewers is clearly confirmed through the increased water use during half-time and at the end of the match, despite the fact that the first match was played in the middle of the night (with different time zones between the Netherlands and the USA). Both graphs point almost precisely to the start of the TV broadcast at 01:50 and 18:50, respectively. The water demand dropped soon after the start of the match until half-time when the first peak occurs; it is not difficult to guess for what purpose the water was used! The upper graphs in both figures show the demand under normal conditions, one week before the game during the same period of the day.

The same phenomenon was observed four years later, in July 1998, during the knockout phase of the World Cup in France, in which the Dutch were playing against Brazil. Figure 2.15 shows an even more exciting game entering into extra time and penalties, and the peak water use in literally every break of the game. This 'TV football-driven consumption' is not only encountered in the Netherlands but virtually everywhere where football is sufficiently popular. Its consequence is a temporary drop in pressure in the system while in the most extreme situations several pumps may automatically shut-off due to excessive flows. Yet, these demand peaks are rarely considered as design parameters because a well-timed adjustment of the pump settings can easily solve the problem.

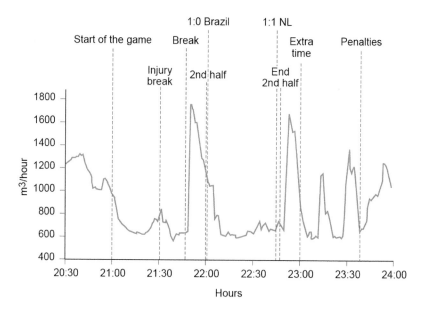

Figure 2.15 Water demand during the football match Brazil vs. NL - World Cup 1998
Source: Water Company 'N-W Brabant', NL, 1998.

Problem 2-2

In a residential area of 10,000 inhabitants, the specific water demand is estimated at 100 l/c/d (leakage included). During a football game shown on the local TV station, the water meter in the area registered a maximum flow of 24 l/s, which was 60 % above the regular use for that period of the day. What was the instantaneous peak factor in that case? What would be the regular peak factor on a day without a televised football broadcast?

Answers

In order to calculate the peak factors, the average demand in the area has to be converted to the same units as the peak flows. Thus, the average flow becomes:

$$Q_{avg} = \frac{10,000 \times 100}{24 \times 3600} \approx 12 \; l/s$$

The regular peak flow at a particular point of the day is 60 % lower than the one registered during the football game, which is 24/1.6 = 15 l/s. Finally, the corresponding peak factors will be 24/12 = 2 during the football game, and 15/12 = 1.25 in normal supply situations.

Self-study: Workshop problems 1.1.3 to 1.1.5 (Appendix 1)

2.3.2   Diurnal patterns

For sufficiently large groups of consumers, the instantaneous demand pattern for 24-hour period converts into a *diurnal (daily) demand diagram*. Diurnal diagrams are important for the design of primary and secondary networks, and in particular their reservoirs and pumping stations. Being the shortest cycle of water use, a one-day period implies a synchronised operation of the system components with similar supply conditions occurring every 24 hours.

Diurnal demand diagram

The diurnal demand patterns are usually registered by monitoring flows at delivery points (treatment plants) or points in the network (pressure boosting stations, reservoirs, and control points with either permanent or temporary measuring equipment). Figure 2.16 shows the diurnal pattern in the city of Rotterdam observed in one of the water treatment plants supplying this city in July 2007.

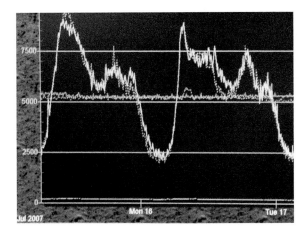

Figure 2.16 Diurnal demand pattern in Rotterdam

Courtesy: Water Company 'Evides', NL, 2007.

With properly organised measurements the diurnal patterns can also be observed at the consumers' premises. First, such an approach allows the separation of various consumption categories and second, the leakage in the distribution system will be excluded, resulting in a genuine consumption pattern.

Some examples of diagrams for different daily demand categories are given in figures 2.17 to 2.20.

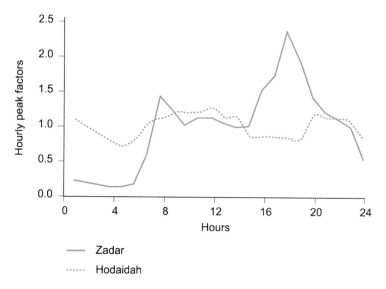

Figure 2.17 Urban demand pattern in two cities

Source: Gabrić, 1996 and Trifunović, 1993.

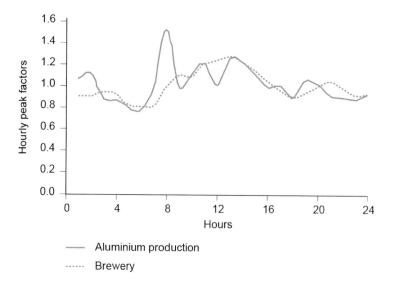

Figure 2.18 Industrial demand pattern - example from Bosnia and Herzegovina

Source: Obradović, 1991.

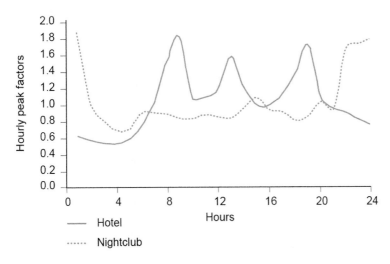

Figure 2.19 Tourist demand pattern - example from Croatia

Source: Obradović, 1991.

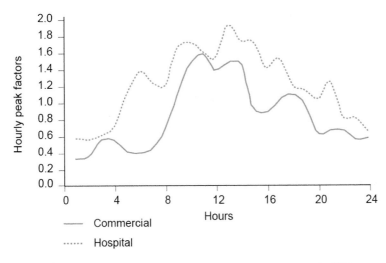

Figure 2.20 Commercial/institutional demand pattern - example from the USA

Source: Obradović, 1991.

A flat daily demand pattern reflects the combination of impacts from the following factors:

- large distribution area,
- high industrial demand,
- high leakage level, and
- scarce supply (individual storage).

Commonly, the structure of the demand pattern in urban areas looks like the one shown in Figure 2.21: the domestic category will have the most visible variation of consumption throughout the day, industry and institutions will usually work in daily shifts, and the remaining categories, including leakage, are practically constant.

By separating the categories, the graph will look like Figure 2.22, with peak factors calculated for the domestic consumption only, then for the total consumption (excluding leakage), and finally for the total demand (consumption plus leakage). It clearly shows that contributions from the industrial consumption and leakage flatten the patterns.

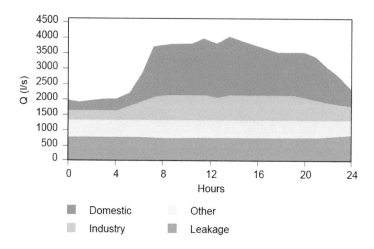

Figure 2.21 Typical structure of diurnal demand in urban areas

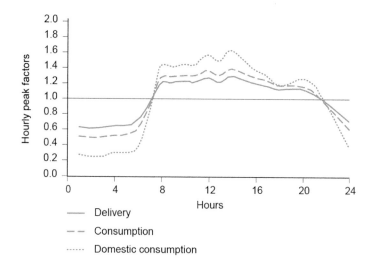

Figure 2.22 Peak factor diagrams of various categories from Figure 2.21

2.3.3 Periodic variations

The peak factors from diurnal diagrams are derived on the basis of average consumption during 24 hours. This average is subject to two additional cycles: weekly and annual.

Weekly demand pattern is influenced by average consumption on working and non-working days. Public holidays, sports events, weather conditions, etc. play a role as well. One example of the demand variations during a week is shown in Figure 2.23. The difference between the two curves in this diagram reflects the successful implementation of the leak detection programme.

Weekly demand pattern

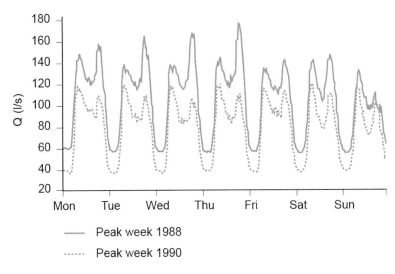

Figure 2.23 Weekly demand variations - Alvington, UK

Source: Dovey and Rogers, 1993.

Consumption in urban areas of Western Europe is normally lower over weekends. On Saturdays and Sundays people rest, which may differ in other parts of the world. For instance, Friday is a non-working day in Islamic countries and domestic consumption usually increases then.

Annual variations in water use are predominantly linked to the change of seasons and are therefore also called *seasonal variations*. The unit consumption per capita normally grows during hot seasons but the increase in total demand may also result from a temporarily increased number of consumers, which is typical for holiday resorts. Figure 2.24 shows the annual pattern in four towns in Istria, Croatia in 2015; the peaks in the tourist season, during July and August, are also the peaks in water use.

Seasonal variations

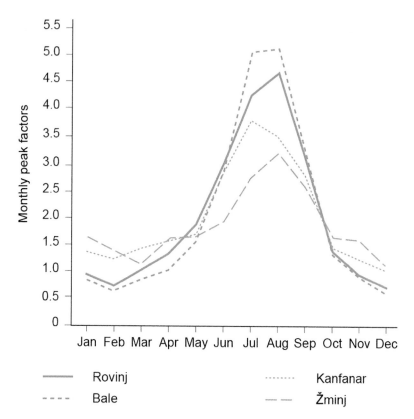

Figure 2.24 Seasonal demand variation in Istria in 2015

Source: Istria Water Supply Company.

Both Bale and specifically Rovinj which is a seaside resort have a larger number of registered tourists in the summer time than the other two towns. This is also reflected in the wider range of seasonal peak factors influenced by the tourist numbers on top of what would be the increase in specific consumption due to the increase of ambient temperature. Figure 2.25 also shows a clear correlation between the monthly water demand, the average monthly temperature and the number of registered tourist nights in Rovinj in 2015.

Just as with diurnal patterns, typical weekly and annual patterns can also be expressed through peak factor diagrams. Figure 2.26 shows an example in which the peak daily demand appears typically on Mondays and is 14 % above the average, while the minimum on Sundays is 14 % below the average daily demand for the week. The second curve shows the difference in demand between summer and winter months, fluctuating within a margin of 10 %.

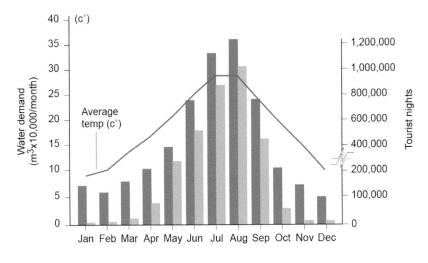

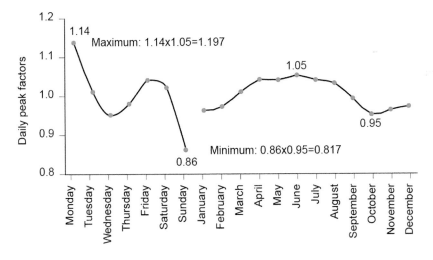

Figure 2.25 Monthly demand, ambient temperatures and tourist nights in the seaside resort of Rovinj in Istria, Croatia

Source: Istria Water Supply Company, 2015; World Climate Guide, 2016; Istria Tourist Board, 2015.

Figure 2.26 Weekly and monthly peak factor diagrams

Generalising such trends leads to the conclusion that the absolute peak consumption during one year occurs on a day of the week, and in the month when the consumption is statistically the highest. This day is commonly called the *maximum consumption day*. In the above example, the maximum consumption day would be a Monday somewhere in June, with its consumption being 1.14 × 1.05 ≈ 1.2 times

Maximum consumption day

higher than the average daily consumption for the year. In practice however, the maximum consumption day in one distribution area will be determined from the daily demand records of the water company. This is simply the day when the total registered demand was the highest in a particular year.

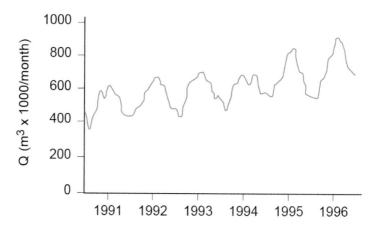

Figure 2.27 Annual demand patterns in Ramallah, Palestinian Authority

Source: Abu-Thaher, 1998.

Finally, the daily, weekly and annual cycles are never repeated in exactly the same way. However, for design purposes a sufficient accuracy is achieved if it is assumed that all water needs are satisfied in a similar schedule during one day, week or year. Regarding the seasonal variations, the example in Figure 2.27 confirms this; the annual patterns in the graph are more or less the same while the average demand grows each year as a result of population growth.

Problem 2-3

A water supply company delivered an annual quantity of 10,000,000 m³, assuming an average leakage of 20 %. On the maximum consumption day, the registered delivery was as follows:

Hour	1	2	3	4	5	6	7	8	9	10	11	12
m³	989	945	902	727	844	1164	1571	1600	1775	1964	2066	2110

Hour	13	14	15	16	17	18	19	20	21	22	23	24
m³	1600	1309	1091	945	1062	1455	1745	2139	2110	2037	1746	1018

Determine:
a) diurnal peak factors for the area,
b) the maximum seasonal variation factor, and
c) diurnal consumption factors.

Answers:

a) From the above table, the average delivery on the maximum con-
    sumption day was 1454.75 m³/h leading to the following hourly peak
    factors:

Hour	1	2	3	4	5	6	7	8	9	10	11	12
pf_h	0.680	0.650	0.620	0.500	0.580	0.800	1.080	1.100	1.220	1.350	1.420	1.450

Hour	13	14	15	16	17	18	19	20	21	22	23	24
pf_h	1.100	0.900	0.750	0.650	0.730	1.000	1.200	1.470	1.450	1.400	1.200	0.700

b) The average delivery, based on the annual figure, is 10,000,000 / 365 /
    24 = 1141.55 m³/h. The seasonal variation factor is therefore 1454.75 /
    1141.55 = 1.274.

c) The average leakage of 20 % assumes an hourly flow of approximately
    228 m³/h, whichis included in the above hourly flows as water loss. The
    peak factors for consumption will therefore be recalculated without this
    figure, as the following table shows:

Hour	1	2	3	4	5	6	7	8	9	10	11	12
m³	761	717	674	499	616	936	1343	1372	1547	1736	1838	1882
pf_h	0.620	0.584	0.549	0.407	0.502	0.763	1.095	1.118	1.261	1.415	1.498	1.534

Hour	13	14	15	16	17	18	19	20	21	22	23	24
m³	1372	1081	863	717	834	1227	1517	1911	1882	1809	1518	790
pf_h	1.118	0.881	0.703	0.584	0.680	1.000	1.237	1.558	1.534	1.475	1.237	0.644

Figure 2.28 shows what the hourly peak factors for the two situations will
look like.

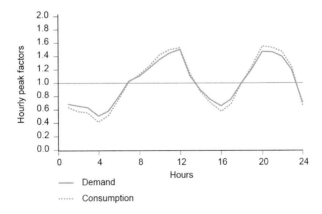

Figure 2.28 Diurnal demand patterns, problems 2-3a and 2-3c

Self-study: Workshop problems 1.1.6 to 1.1.8 (Appendix 1)
            Spreadsheet Lessons 8-2 to 8-4 (Appendix 7)

2.4   Demand calculation

Knowing the daily patterns and periodical variations, the demand flow can be calculated from the following formula:

$$Q_d = \frac{Q_{wc,avg} \times pf_o}{\left(1 - \dfrac{l}{100}\right)f_c} \tag{2.1}$$

The definition of the parameters is as follows:

Q_d = Water demand of a certain area at a certain moment.

$Q_{wc,avg}$ = Average water consumption in the area.

pf_o = Overall peak factor; this is a combination of the peak factor values from the daily, weekly and annual diagrams: $pf_o = pf_h \times pf_d \times pf_m$. The daily and monthly peak factors are normally integrated into one (seasonal) peak factor: $pf_s = pf_d \times pf_m$.

l = Leakage expressed as a percentage of the water production.

f_c = Unit conversion factor.

The main advantage of Equation 2.1 is its simplicity although some inaccuracy will be necessarily introduced. Using this formula, the volume of leakage increases in line with higher consumption i.e. the peak factor value, despite the fixed leakage percentage. For example, if $Q_{wc,avg} = 1$ (regardless of the flow units), $pf_o = 1$ and the leakage percentage is 50 %, then as a result $Q_d = 2$. Thus, half of the supply is consumed and the other half is leaked.

If $pf_o = 2$, $Q_d = 4$. Again, this is 'fifty-fifty' but this time the volume of leakage has grown from 1 to 2, which implies its dependence on the consumption level. This is not true as the leakage level is usually constant throughout the day, with a slight increase overnight when the pressures in the network are generally higher (already visible in Figure 2.21). Hence, *the leakage level is pressure-dependent rather than consumption-dependant.*

Nonetheless, the above inaccuracy effectively adds safety to the design. Where this is deemed unnecessary, an alternative approach is suggested, especially for distribution areas with high leakage percentages:

$$Q_d = \left(Q_{wc,avg} \times pf_o + Q_{wl}\right)\frac{1}{f_c} \tag{2.2}$$

where:

$$Q_{wl} = \frac{l}{100}Q_{wp} \tag{2.3}$$

In the case of $pf_o = 1$, demand equals production and assuming the same units for all parameters ($f_c = 1$):

$$Q_{wp} = Q_{wc,avg} + \frac{l}{100}Q_{wp} \tag{2.4}$$

This can be re-written as:

$$Q_{wp} = \frac{Q_{wc,avg}}{\left(1 - \frac{l}{100}\right)} \tag{2.5}$$

By plugging Equation 2.5 into 2.3 and then to 2.2, the formula for the water demand calculation evolves into its final form:

$$Q_d = \frac{Q_{wc,avg}}{f_c}\left(pf_o + \frac{l}{100 - l}\right) \tag{2.6}$$

Where reliable information resulting from individual metering of consumers is not available, the average water consumption, $Q_{wc,avg}$, can be approximated in several ways:

$$Q_{wc,avg} = ncq \tag{2.7}$$

$$Q_{wc,avg} = dAcq \tag{2.8}$$

$$Q_{wc,avg} = Acq_a \tag{2.9}$$

$$Q_{wc,avg} = ncq \tag{2.10}$$

where:

n = Number of inhabitants in the distribution area.

c = Coverage of the area. It can happen that some of the inhabitants are not connected to the system, or some parts of the area are not inhabited. This factor, which has a value of between 0 and 1, converts the number of inhabitants into the number of consumers.

q = Specific consumption (l/c/d).

d = Population density (number of inhabitants per unit surface area).

A = Surface area of the distribution area.

q_a = Consumption registered per unit surface area.

n_u = Production capacity. It represents a number of units (kg, l, pieces, etc.) produced within a certain period.

q_u = Water consumption per unit product.

The unit consumptions q, q_a, and q_u are elaborated in Section 2.2. The data for n, c, d, A and n_u are usually available from statistics or set by planning: local, urban, regional, etc.

As already mentioned, the demand in large urban areas is often composed of several consumption categories. More accuracy in the calculation of demand for water is therefore achieved if the distribution area is split into a number of sub-areas or districts, with standardised categories of water users and a range of consumptions based on local experience. The average consumption per district can then be calculated from Equation 2.9, which has thus been modified:

$$Q_{wc,avg} = A\sum_{i=1}^{n}\left(q_{a,i}p_ic_i\right) \tag{2.11}$$

where:

A = Surface area of the district.

n = Number of consumption categories within the district.

$q_{a,i}$= Unit consumption per surface area of category i.

p_i = Percentage of the district territory occupied by category i.

c_i = Coverage within the district territory occupied by category i.

With a known population density in each district, the result can be converted into specific demand (per capita).

Regarding the pf_o values, the following are typical combinations:

1) $pf_h = 1$, $pf_s = 1$; Q_d represents the average consumption per day. This demand is the absolute average, usually obtained from annual demand records and converted into required flow units.

2) $pf_h = 1$, $pf_s = $ max; Q_d represents the average demand during the maximum consumption day.

3) $pf_h = $ max, $pf_s = $ max; Q_d is the demand at the maximum consumption hour on the maximum consumption day.

4) $pf_h = $ min, $pf_s = $ min; Q_d is the demand at the minimum consumption hour on the minimum consumption day.

The entire range of demands that appear in one distribution system during one year is specified by the demands shown in Figure 2.29. These peak demands are relevant as parameters for the design of all the system components: pipes, pumps and storage.

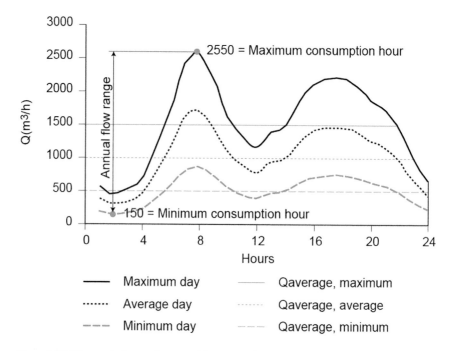

Figure 2.29 Theoretical annual range of flows in a distribution system

Problem 2-4

A water supply company in a town with a total population of approximately 275,000 conducted a water demand survey resulting in the following categories of water users:

Category of water users	q_a (m³/d/ha)
A - Residential area, apartments	90
B - Residential area, individual houses	55
C - Shopping areas	125
D - Offices	80
E - Schools,colleges	100
F - Hospitals	160
G - Hotels	150
H - Public green areas	15

The city is divided into eight districts, each with a known population and contribution to demand from each of the categories, and estimated coverage by the distribution system, as shown in the table below.

Districts		A	B	C	D	E	F	G	H
86,251	$p_1(\%)$	37	23	10	0	4	0	0	26
$A_1=250$ ha	$c_1(\%)$	100	100	100	0	100	0	0	40
74,261	$p_2(\%)$	20	5	28	11	12	0	5	19
$A_2=185$ ha	$c_2(\%)$	100	100	95	100	100	0	100	80
18,542	$p_3(\%)$	10	18	3	0	0	42	0	27
$A_3=57$ ha	$c_3(\%)$	100	100	100	0	0	100	0	35
42,149	$p_4(\%)$	25	28	20	2	15	0	0	10
$A_4=88$ ha	$c_4(\%)$	100	100	95	100	100	0	0	36
22,156	$p_5(\%)$	50	0	11	0	10	0	0	29
$A_5=54$ ha	$c_5(\%)$	100	0	100	0	100	0	0	65
9958	$p_6(\%)$	24	11	13	15	13	8	0	16
$A_6=29$ ha	$c_6(\%)$	100	100	100	100	100	100	0	35
8517	$p_7(\%)$	22	28	8	19	6	0	10	7
$A_7=17$ ha	$c_7(\%)$	100	100	100	100	100	0	100	50
12,560	$p_8(\%)$	0	0	0	0	55	20	15	10
$A_8=16$ ha	$c_8(\%)$	0	0	0	0	85	100	100	45

Determine the total average demand of the city.

Answer:

Based on Formula 2.11, a sample calculation of the demand for District 1 will be as follows:

$$Q_{1,avg} = A\sum_{i=1}^{5}\left(q_{a,i}p_i c_i\right)$$
$$= 250\times(90\times0.37+55\times0.23+125\times0.10+100\times0.04+15\times0.26\times0.40)$$

The above calculation yields the demand of 666.77 m³/h.

The remainder of the results are shown in the table below. The specific demand has been calculated based on the registered population in each district.

	Q_{avg} (m³/h)	Population	Q_{avg} (l/c/d)
District 1	666.67	86,251	186
District 2	651.97	74,261	211
District 3	216.76	18,542	281
District 4	288.90	42,149	165
District 5	161.05	22,156	174
District 6	99.74	9958	240
District 7	58.03	8517	164
District 8	67.95	12,560	130
TOTAL	2211.16	274,394	193

Self-study: Spreadsheet Lessons 8-5 to 8-7 (Appendix 7)

2.5   Demand forecasting

Water demand usually grows unpredictably because it depends on many parameters that all have their own unpredictable trends. Figure 2.30 illustrates how the rate of increase in consumption may differ even in countries from the same region and with a similar level of economic development.

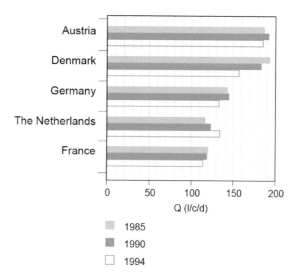

Figure 2.30 Increase in domestic consumption in a number of European countries
Source: EEA, 2001.

The experience from Germany proves again how uncertain the forecast can be. Figure 2.31 shows the development of domestic consumption in the period 1970- 2000. The forecasts made in the period 1972-1980 were that the demand in the year 2000 would grow to as high as 220 l/c/d, while in reality it was approximately 140 l/c/d.

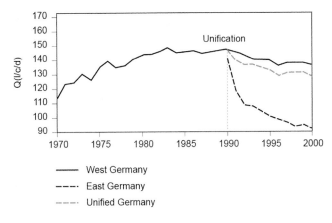

Figure 2.31 Increase in domestic consumption in Germany

Source: BGW.

Awareness about the environment in the last few decades, combined with low population growth, has caused a drop in domestic water use in many countries of Western Europe. In addition, lots of home appliances (i.e. shower heads, taps, washing machines, dishwashers, etc.) have been replaced with more advanced models, able to achieve the same effect with less water. The example in Figure 2.32 illustrates the development of technology in the production of washing machines and dishwashers, which has managed to reduce the needed quantity of water per washing cycle to less than a third (EEA, 2001).

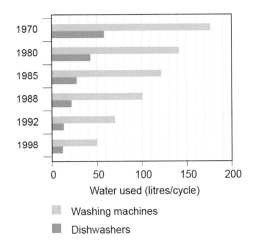

Figure 2.32 Water consumption by washing appliances in Europe

Source: EEA, 2001.

This is unfortunately less the case in many developing countries and countries in transition where the population growth is much faster, consumers have less interest in water conservation, and outdated technologies and equipment are still widely used due to lower living standards. Nevertheless, the trend of a reduction in specific demand is starting to appear there as well, which is also being achieved through more realistic water pricing, increased metering coverage, and reduction of physical losses in the distribution network. Figure 2.33 shows such a result in Belgrade, Serbia (BWS, 2015).

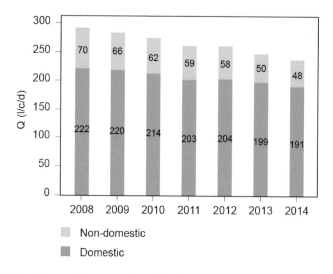

Figure 2.33 Trends in specific water demand in Belgrade, Serbia

Source: BWS, 2015.

Moreover, in order to stay competitive, the manufacturers have increasingly undertaken pro-active marketing of the water-saving features of their products, as the example in Figure 2.34 shows.

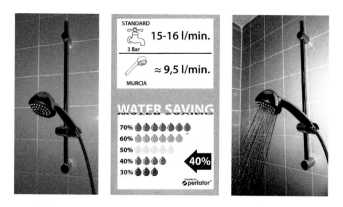

Figure 2.34 Packaging indicating the water-saving capacity of a shower

Manufacturer: Tiger, model: Murcia ECO Chroom.

Apart from monitoring technological developments, several other assessments must be taken into account while estimating future demand:

- historical demand growth patterns,
- projections based on per-capita consumption and population growth trends for the domestic category,
- forecast based on an assessment of growth trends of other main consumer categories (industry, tourism, etc.).

When combined, all these assessments can yield several possible scenarios for consumption growth. While thinking, for instance, about population growth, which is for many developing countries still the major factor in the increase in water demand, useful conclusions can be drawn if the composition of the existing population, fertility and mortality rates, and particularly the rate of migration, can be assessed. That the population and demand growth match reasonably well in general is shown by the example in Figure 2.35.

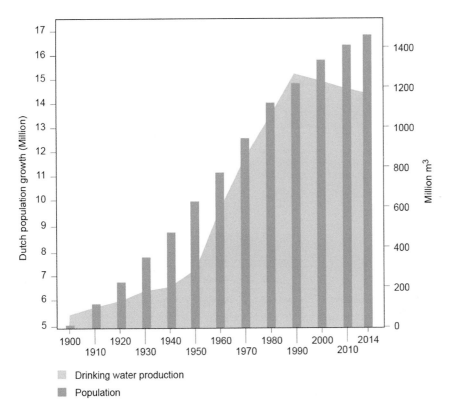

Figure 2.35 Population and demand growth in the Netherlands

Source: VEWIN; CBS.

Two models are commonly used for demand forecast: linear and exponential.

Linear model

$$Q_{i+n} = Q_i \left(1 + n\frac{a}{100}\right)$$ (2.12)

Exponential model

$$Q_{i+n} = Q_i \left(1 + \frac{a}{100}\right)^n$$ (2.13)

In the above equations:
Q_i = Water demand at year i.
Q_{i+n} = Forecast water demand after n years.
n   = Design period.
a   = Average annual population growth during the design period (%).

Which of the models will be more suitable will depend on the conclusions from the above-mentioned analyses. These should be reviewed periodically, as trends can change within a matter of years.

Figure 2.36 shows the annual demand in the Netherlands in the period 1955-2014. In the first part of this period, up until 1970, the exponential model with an average annual growth of 5.5 % (a in Equation 2.13 = 5.5) matches the real demand very closely. However, keeping it unchanged for the entire period would show demand almost ten times higher than was actually registered in 2014!

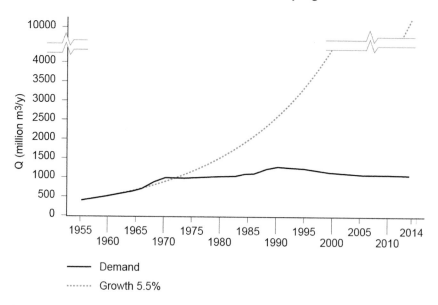

Figure 2.36 Demand growth in the Netherlands according to the exponential model

The actual growth trend of domestic and non-domestic demand in the Netherlands in the period 1960-2013 is shown in Figure 2.37.

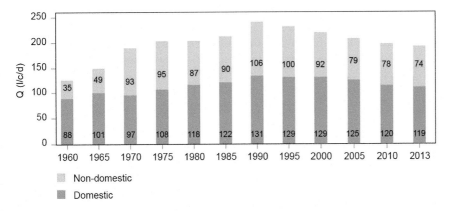

Figure 2.37 Growth trends in domestic and non-domestic demand in the Netherlands
Source: VEWIN.

Problem 2-5

In a residential area of 250,000 inhabitants, the specific water demand is estimated at 150 l/c/d, which includes leakage. Calculate the demand in 20 years'
time if the assumed annual demand growth is 2.5 %. Compare the results by
applying the linear and exponential models.

Answers:

The present demand in the city is equal to:

$$Q_{avg} = \frac{250,000 \times 150 \times 365}{1000} = 91,250 \ m^3 \ / \ y$$

Applying the linear model, the demand after 20 years will grow to:

$$Q_{21} = 91,250 \times \left(1 + 20\frac{2.5}{100}\right) = 136,875 \ m^3 \ / \ y$$

which is an increase of 50 % compared to the present demand. In the case of
the exponential model:

$$Q_{21} = 91,250 \times \left(1 + \frac{2.5}{100}\right)^{20} \approx 149,524 \ m^3 \ / \ y$$

which is an increase of approximately 64 % compared to the present demand.

Self-study: Workshop problems 1.1.9 and 1.1.10 (Appendix 1)
                Spreadsheet Lesson 8-8 (Appendix 7)

2.6   Demand frequency distribution

A water supply system is generally designed to satisfy the demand at guaranteed minimum pressures, for 24 hours a day and 365 days per year. Nevertheless, if the pressure threshold is set high, such a level of service may require exorbitant investment that is actually non-affordable for the water supply company and consumers. It is therefore useful to analyse how often the maximum peak demands occur during the year. The following example explains the principle.

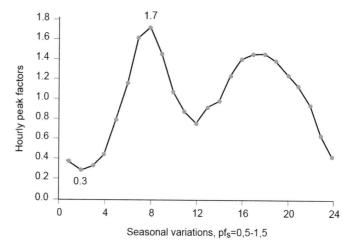

Figure 2.38 Example of a typical diurnal demand pattern

Knowing both a typical diurnal peak factor diagram (such as the one shown in Figure 2.38) and the range of seasonal peak factors allows for integration of the two. The hourly peak factors corrected by the seasonal peak factors will result in the annual range of the hourly peak hours (Figure 2.39).

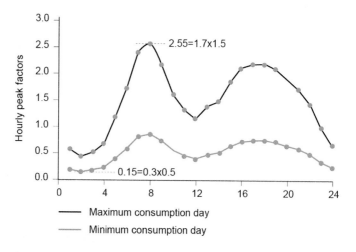

Figure 2.39 Example of the annual range of the peak factors

These are absolute values that refer to the average hour of the average consumption day, which is 1000 m³/h in the example in Figure 2.40. Consequently, each hour of the year (total 24 × 365) will have a unique peak factor value assigned to it.

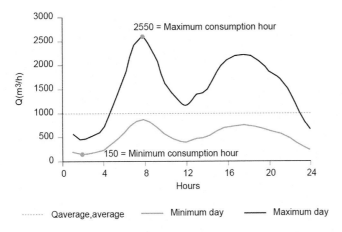

Figure 2.40 Example of the annual range of hourly demands

Applying this logic, a diagram with the frequency distribution of all the hourly peak factors can be plotted, as the example in Figure 2.41 shows. Converting this diagram into a cumulative frequency distribution curve helps to determine the number of hours in the year when the peak factors exceed the corresponding value. This is illustrated in Figure 2.42, which for instance shows that the peak factors above 2.0 only appear during some 500 hours or approximately 5 % of the year. In theory, excluding this fraction from the design considerations would eventually create savings based on a 20 % reduction of the system capacity. The consequence of such a choice would be the occasional drop of pressure below the threshold, which the consumers might consider acceptable for a limited period of time.

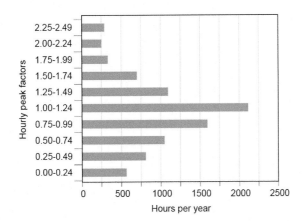

Figure 2.41 Frequency distribution of the diurnal peak factors

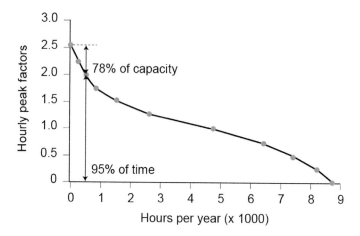

Figure 2.42 Cumulative frequency distribution of the diurnal peak factors

In practice, the decision about the design peak factor is based on a comparison of the costs and benefits. Indeed, it seems rather inefficient to lay pipes that will be used for 80 % of their capacity for less than 5 % of the total time. However, where there is a considerable scope for energy savings in daily use by lowering the energy losses on a wider scale, such a choice may look reasonable. Moreover, careful assessment of the network reliability could justify the laying of pipes with reserve capacity that could be utilised during irregular supply situations. Finally, some spare capacity is also useful for practical reasons since it can postpone the construction of phased extensions to expand the system.

The relation between demand for water and hydraulic losses is thoroughly discussed in the following chapter.

Self-study: Spreadsheet Lesson 8-9 (Appendix 7)

Chapter 3

Steady flow in pressurised networks

3.1 Main concepts and definitions

The hydraulic principles applied in water transport and distribution practice are based on three main assumptions:

1) the system is filled with water under pressure,
2) that water is incompressible, and
3) that water has a steady and uniform flow.

In addition, it is assumed that distortion of the system boundaries is negligible, meaning that the water flows through a non-elastic system[1]

Flow Q (m³/s) through a pipe cross section of area A (m²) is determined as Q = v × A, where v (m/s) is the mean velocity in the cross section. This flow is *steady* if the mean velocity remains constant over a period of time Δt. *Steady flow*

If the mean velocities of two consecutive cross sections are equal at a particular moment in time, the flow between the cross sections is *uniform*. *Uniform flow*

The above definitions written in the form of equations for two moments in time that are close together, t_1 and t_2, and in the cross sections 1 and 2 (Figure 3.1) yield:

$$v_1^{(t_1)} = v_1^{(t_2)} \wedge v_2^{(t_1)} = v_2^{(t_2)} \tag{3.1}$$

for a steady flow, and:

$$v_1^{(t_1)} = v_2^{(t_1)} \wedge v_1^{(t_2)} = v_2^{(t_2)} \tag{3.2}$$

for a uniform flow.

A steady flow in a pipe with a constant diameter is also uniform. Thus:

$$v_1^{(t_1)} = v_2^{(t_1)} = v_1^{(t_2)} = v_2^{(t_2)} \tag{3.3}$$

These simplifications help to describe the general hydraulic behaviour of water distribution systems assuming that the time interval between t_1 and t_2 is sufficiently short. Relatively slow changes in boundary conditions during regular operation of these systems make Δt of a few minutes acceptably short for the assumptions introduced above. This interval is also long enough to simulate changes in the

---

1 The foundations of steady-state hydraulics are described in detail in various references to fluid mechanics and engineering hydraulics. See for instance Streeter and Wylie (1985).

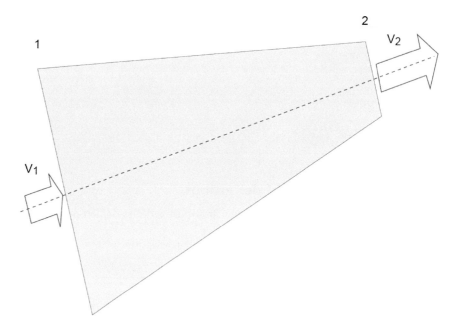

Figure 3.1 Velocities in consecutive cross sections

pump operation, levels in the reservoirs, diurnal demand patterns, etc., without handling unnecessarily large amounts of data.

If there is a sudden change in operation, for instance a situation caused by pump failure or valve closure, transitional flow conditions occur in which the assumptions of the steady and uniform flow are no longer valid. To be able to accurately describe these phenomena mathematically, a more complex approach elaborated on in the theory of *transient flows* would be required, which is not discussed in this book. The reference literature on this topic includes Larock *et al.* (2000).

Transient flow

3.1.1   Conservation laws

The conservation laws of mass, energy and momentum are three fundamental laws related to fluid flow. These laws state:

1)   *The Mass Conservation Law*
     Mass m (kg) can neither be created nor destroyed; any mass that enters a system must either accumulate in that system or leave it.
2)   *The Energy Conservation Law*
     Energy E (J) can neither be created nor destroyed; it can only be transformed into another form.
3)   *The Momentum Conservation Law*
     The sum of external forces acting on a fluid system equals the change of the momentum rate M (N) of that system.

These conservation laws are translated into practice through the application of three equations, respectively:

1) the Continuity Equation,
2) the Energy Equation,
3) the Momentum Equation.

The *Continuity Equation* is used when balancing the volumes and flows in distribution networks. Assuming that water is an incompressible fluid, i.e. with a mass density $\rho = m/V = const$ (expressed in kg/m³), the Mass Conservation Law can be applied to volumes. In this situation, the following is valid for tanks: *Continuity Equation*

$$Q_{inp} = Q_{out} \pm \frac{\Delta V}{\Delta t} \qquad (3.4)$$

where $\Delta V/\Delta t$ represents the change in volume V (m³) within a time interval Δt (s). Thus, the difference between the input and output flows from a tank is the volume that is:

1) accumulated in the tank if $Q_{out} < Q_{inp}$ (the + sign in Equation 3.4; the situation illustrated in Figure 3.2),
2) withdrawn from the tank if $Q_{out} > Q_{inp}$ (the – sign in Equation 3.4).

Applied at node n which connects j pipes, the Continuity Equation can be written as:

$$\sum_{i=1}^{j} Q_i - Q_n = 0 \qquad (3.5)$$

where Q_n represents the nodal discharge. An example of three pipes and a discharge point is shown in Figure 3.3.

The *Energy Equation* establishes the energy balance between any two cross sections of a pipe: *Energy Equation*

$$E_1 = E_2 \pm \Delta E \qquad (3.6)$$

where ΔE is the amount of transformed energy between cross sections 1 and 2. It is usually the energy lost from the system (the + sign in Equation 3.6), but may also be added energy caused by pumping of water (the – sign).

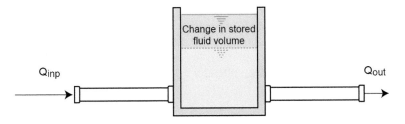

Q_inp   Change in stored
          fluid volume   Q_out

Figure 3.2 The Continuity Equation validation in tanks

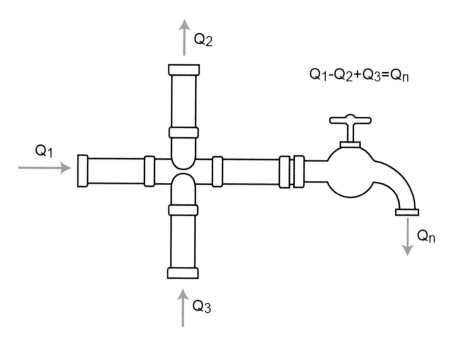

Figure 3.3 The Continuity Equation validation in pipe junctions

Momentum Equation

The *Momentum Equation* (also known as the *Dynamic Equation*) describes the pipe resistance to dynamic forces caused by the pressurised flow. For incompressible fluids, the momentum M (N) carried across a pipe section is defined as:

$$M = \rho Q v \tag{3.7}$$

where ρ (kg/m³) represents the mass density of water. Other forces in the equilibrium are (see Figure 3.4):

1) The hydrostatic force F_h (N) caused by fluid pressure p (N/m² or Pa); $F_h = p \times A$, where A is the pipe cross-section area (m²).
2) The weight w (N) of the considered fluid volume (which only acts in a vertical direction).
3) Force F (N) of the solid surface acting on the fluid.

The Momentum Equation as written for a horizontal direction would state:

$$\rho Q v_1 - \rho Q v_2 \cos \phi = -p_1 A_1 + p_2 A_2 \cos \phi + F_x \tag{3.8}$$

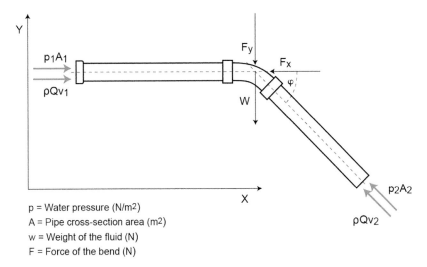

p = Water pressure (N/m2)
A = Pipe cross-section area (m2)
w = Weight of the fluid (N)
F = Force of the bend (N)

Figure 3.4 The Momentum Equation

whereas in a vertical direction:

$$\rho Q v_2 \sin \phi = -p_2 A_2 \sin \phi + w + F_y \qquad (3.9)$$

The forces of the water acting on the pipe bend are the same, i.e. F_x and F_y but with an opposite direction i.e. a negative sign, in which case the total force, known as the *pipe thrust,* will be:

Pipe thrust

$$F = \sqrt{F_x^2 + F_y^2} \qquad (3.10)$$

The Momentum Equation is applied in calculations for the additional strengthening of pipes in locations where the flow needs to be diverted. The results are used for the design of concrete structures required for anchoring of pipe bends and elbows.

Problem 3-1

A velocity of 1.2 m/s has been measured in a pipe of diameter D = 600 mm. Calculate the pipe flow.

Answer:

The cross section of the pipe is:

$$A = \frac{D^2 \pi}{4} = \frac{0.6^2 \times 3.14}{4} = 0.2827 \, m^2$$

which yields the flow of:

$$Q = vA = 1.2 \times 0.2827 = 0.339 \, m^3 \, / \, s \approx 340 \, l \, / \, s$$

Problem 3-2

A circular tank with a diameter at the bottom of D = 20 m and with vertical walls has been filled with a flow of 240 m³/h. What will be the increase in the tank depth after 15 minutes, assuming a constant flow during this period of time?

Answer:

The tank cross section area is:

$$A = \frac{D^2 \pi}{4} = \frac{20^2 \times 3.14}{4} = 314.16 \, m^2$$

The flow of 240 m³/h fills the tank with an additional 60 m³ after 15 minutes, which will increase the tank depth by a further 60 / 314.16 = 0.19 m ≈ 20 cm.

Problem 3-3

For a pipe bend of 45° and a continuous diameter of D = 300 mm, calculate the pipe thrust if the water pressure in the bend is 100 kPa at a measured flow rate of 26 l/s. The weight of the fluid can be neglected. The mass density of the water equals ρ = 1000 kg/m³.

Answer:

From Figure 3.4, for a continuous pipe diameter:

$$A_1 = A_2 = \frac{D^2 \pi}{4} = \frac{0.3^2 \times 3.14}{4} = 0.07 \, m^2$$

Consequently, the flow velocity in the bend can be calculated as:

$$v_1 = v_2 = \frac{Q}{A} = \frac{0.026}{0.07} = 0.37 \, m \, / \, s$$

Furthermore, for the angle φ = 45°, sin φ = cos φ = 0.71. Assuming also that p_1 = p_2 = 100 kPa (or 100,000 N/m²), the thrust force in the X direction becomes:

$$-F_x = 0.29 \times (pA + \rho Q v) = 0.29 \times (100,000 \times 0.07 + 1000 \times 0.026 \times 0.37) \approx 2030 \, N = 2 \, kN$$

while in the Y-direction:

$$-F_y = 0.71 \times (pA + \rho Q v) \approx 5 \, kN$$

The total force will therefore be:

$$F = \sqrt{2^2 + 5^2} \approx 5.4 \, kN$$

The calculation shows that the impact of water pressure is much more significant than the impact of flow/velocity.

Self-study: Spreadsheet Lesson 1-1 (Appendix 7)

3.1.2   Energy and hydraulic grade lines

The energy balance in Equation 3.6 stands for total energies in two cross sections of a pipe. The total energy in each cross section comprises three components, which is generally written as:

$$E_{tot} = mgZ + m\frac{p}{\rho} + \frac{mv^2}{2} \tag{3.11}$$

expressed in J or more commonly in kWh. Written per unit weight mg (N), the equation looks as follows:

$$E_{tot} = Z + \frac{p}{\rho g} + \frac{v^2}{2g} \tag{3.12}$$

where the energy obtained will be expressed in *metres of water column* (mwc). Parameter g in both these equations stands for gravity (9.81 m/s²).

The first term in equations 3.11 and 3.12 determines the *potential energy*, which is entirely dependent on the elevation of the mass/volume.   *Potential energy*

The second term stands for the flow energy that comes from the ability of a fluid mass m = ρ × V to do a work W (N) generated by the above-mentioned pressure forces F = p × A. At pipe length L, these forces create the work that can be described per unit mass as:

$$W = FL = \frac{pAL}{\rho V} = \frac{p}{\rho} \tag{3.13}$$

Finally, the third term in the equations represents the *kinetic energy* generated   *Kinetic energy*
by the mass/volume motion.

By plugging 3.12 into 3.6, it becomes:

$$Z_1 + \frac{p_1}{\rho g} + \frac{v_1^2}{2g} = Z_2 + \frac{p_2}{\rho g} + \frac{v_2^2}{2g} \pm \Delta E \tag{3.14}$$

In this form, the energy equation is known as the *Bernoulli Equation*. The   *Bernoulli*
equation parameters are shown in Figure 3.5. The following terminology is in   *Equation*
common use:

- *Elevation head:* $Z_{1(2)}$
- *Pressure head:* $p_{1(2)}/\rho g$
- *Piezometric head:* $H_{1(2)} = Z_{1(2)} + p_{1(2)}/\rho g$

- *Velocity head*: $v^2_{1(2)}/2g$
- *Energy head*: $E_{1(2)} = H_{1(2)} + v^2_{1(2)}/2g$

The pressure head and velocity head are expressed in mwc, which gives a good visual impression while talking about 'high' or 'low' pressures or energies. The elevation head, piezometric head and energy head are compared to a reference or 'zero' level. Any level can be taken as a reference; it is commonly the mean sea level suggesting the units for Z, H and E in *metres above mean sea level* (msl)[2]. Alternatively, the street level can also be taken as a reference. To provide a link with the SI units, the following is valid:

- 1 mwc of the pressure head corresponds to 9.81 kPa in SI units, which for practical reasons is often rounded off to 10 kPa.
- 1 mwc of the potential energy corresponds to 9.81 ($\approx$10) kJ in SI units; for instance, this energy will be possessed by 1 m³ of the water volume elevated 1 m above the reference level.
- 1 mwc of the kinetic energy corresponds to 9.81 ($\approx$10) kJ in SI units; for instance, this energy will be possessed by 1 m³ of the water volume flowing at a velocity of 1 m/s.

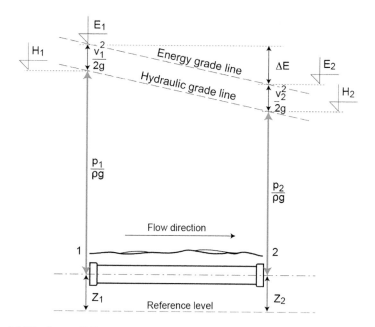

Figure 3.5 The Bernoulli Equation

2 For the sake of simplicity, the pressure in mwc is in other literature commonly expressed in 'meters' alone, without mentioning the 'water column'. The unit of mwc is deliberately applied throughout this book in order to distinguish between the pressures and piezometric/energy heads, the latter being expressed consistently in msl.

In reservoirs with a surface level in contact with the atmosphere, pressure p equals the atmospheric pressure, hence $p = p_{atm} \approx 0$. Furthermore, the velocity throughout the reservoir volume can be neglected ($v \approx 0$ m/s). As a result, both the energy head and piezometric head will be positioned at the surface of the water. Hence, $E_{tot} = H = Z$. Reservoir water depth can alternatively be expressed as a pressure in which case $H - Z = p/\rho g$.

The lines that indicate the energy head level and piezometric head level in consecutive cross sections of a pipe are called the *energy grade line* and the *hydraulic grade line*, respectively.

Energy and hydraulic grade lines

The energy and hydraulic grade lines are parallel for uniform flow conditions. Furthermore, the velocity head is in reality considerably smaller than the pressure head. For example, for a common pipe velocity of 1 m/s, $v^2/2g = 0.05$ mwc, while the pressure heads are often in the order of tens of metres of water column. Hence, the real difference between these two lines is, with a few exceptions, negligible and the hydraulic grade line is usually used when solving practical problems. Its position and slope indicate the flow direction and the pressures existing in the pipe, as can be seen in Figure 3.6. The space between the pipe and the hydraulic grade line gives an indication of the pressure buffer available in the network. As well as being useful to mitigate irregular service, this buffer also boosts water losses during regular supply.

The hydraulic grade line is generally not parallel to the slope of the pipe which normally varies from section to section. In hilly terrains, the energy level may

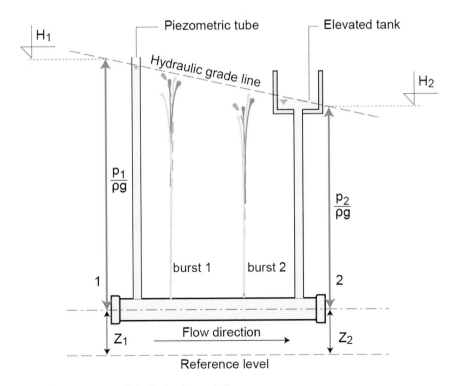

Figure 3.6 Depiction of the hydraulic grade line

Backsiphonage and cross connection

even drop below the pipe invert causing negative pressure (below atmospheric), as Figure 3.7 shows. The negative pressure is known in practice as a *backsiphonage* which needs to be avoided to prevent contamination intrusion through pipe cracks or by a *cross connection* between potable and non-potable part of water distribution system.

Hydraulic gradient

The slope of the hydraulic grade line is called the *hydraulic gradient*, $S = \Delta E/L = \Delta H/L$, where L (m) is the length of the pipe section. This parameter reflects the pipe conveyance (Figure 3.8).

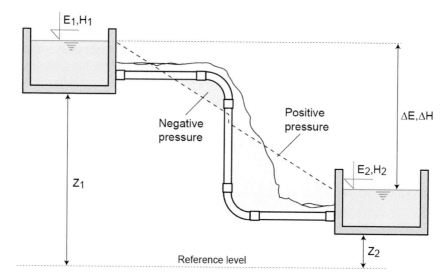

Figure 3.7 The hydraulic grade line

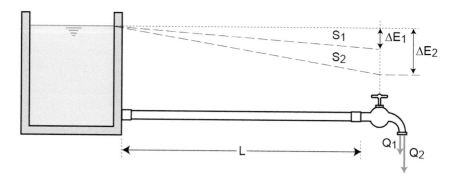

Figure 3.8 The hydraulic gradient

The flow rate in pipes under pressure is related to the hydraulic gradient and not to the slope of the pipe. More energy is needed for a pipe to convey more water, which is expressed in the higher value of the hydraulic gradient.

Problem 3-4

For the pipe bend in Problem 3-3 (Section 3.1.1), calculate the total energy head and piezometric head in the cross section of the bend if it is located at $Z = 158$ msl. Express the result in msl, J and kWh.

Answer:

In Problem 3-3, the pressure indicated in the pipe bend is $p = 100$ kPa, while the velocity, calculated from the flow rate and the pipe diameter, is $v = 0.37$ m/s. The total energy can be determined from Equation 3.12:

$$E_{tot} = Z + \frac{p}{\rho g} + \frac{v^2}{2g} = 158 + \frac{100,000}{1000 \times 9.81} + \frac{0.37^2}{2 \times 9.81}$$
$$= 158 + 10.194 + 0.007 = 168.2 \; msl$$

As can be seen, the impact of the kinetic energy is minimal and the difference between the total energy and the piezometric head can therefore be neglected. The same result in J and kWh is as follows:

$$E_{tot} = 168.2 \times 1000 \times 9.81 = 1,650,042 \, J \approx 1650 \, kJ \; (or \; kWs)$$
$$= \frac{1650}{3600} \approx 0.5 \, kWh$$

For an unspecified volume, the above result represents a type of unit energy, expressed per m³ of water. Remember the unit conversion: $1 \, N = 1 \, kg \times m/s^2$ and $1 \, J = 1 \, N \times m$.

3.2 Hydraulic losses

The energy loss ΔE from Equation 3.14 is generated by:

- friction between the water and the pipe wall, and
- turbulence caused by obstructions of the flow.

These causes inflict the *friction loss* and *minor loss*, respectively. Both can be expressed in the same format:

$$\Delta E = h_f + h_m = R_f Q^{n_f} + R_m Q^{n_m} \tag{3.15}$$

where R_f stands for *resistance* of a pipe with diameter D, along its length L.   *Pipe resistance*
The parameter R_m can be characterised as a resistance at the pipe cross section where an obstruction occurs. Exponents n_f and n_m depend on the type of equation applied.

3.2.1   Friction losses

The most popular equations used for the determination of friction losses are:

1) the Darcy-Weisbach Equation,
2) the Hazen-Williams Equation,
3) the Manning Equation.

Following the format in Equation 3.15:

Darcy-Weisbach

$$R_f = \frac{8\lambda L}{\pi^2 g D^5} = \frac{\lambda L}{12.1 D^5}; \; n_f = 2 \tag{3.16}$$

Hazen-Williams

$$R_f = \frac{10.68 L}{C_{hw}^{1.852} D^{4.87}}; \; n_f = 1.852 \tag{3.17}$$

Manning

$$R_f = \frac{10.29 N^2 L}{D^{16/3}}; \; n_f = 2 \tag{3.18}$$

In all three cases, the friction loss h_f will be calculated in mwc for the flow Q expressed in m³/s and for length L and diameter D expressed in m. The Darcy-Weisbach and Manning equations are dimensionally homogeneous, which is not the case with the Hazen-Williams Equation. Nevertheless, *it is essential to use the prescribed parameter units in equations 3.16-3.18 because the constants will need to be readjusted depending on the alternative units used.*
In the above equations, λ, C_{hw} and N are experimentally-determined factors that describe the impact of the pipe wall roughness on the friction loss.

The Darcy-Weisbach Equation

Colebrook-White Equation

In the Darcy-Weisbach Equation, the friction factor λ (-) (also labelled as f in some literature) can be calculated from the *Colebrook-White Equation*:

$$\frac{1}{\sqrt{\lambda}} = -2\log\left[\frac{2.51}{Re\sqrt{\lambda}} + \frac{k}{3.7D}\right] \tag{3.19}$$

where:
k = Absolute roughness of the pipe wall (mm).
D = Inner diameter of the pipe (mm).
Re = Reynolds number (-).

To avoid iterative calculation, *Barr* (1975) suggests the following acceptable approximation, which has an average accuracy of $\pm 1\%$:

$$\frac{1}{\sqrt{\lambda}} = -2\log\left[\frac{5.1286}{\mathrm{Re}^{0.89}} + \frac{k}{3.7D}\right] \tag{3.20}$$

A few other approximations resulting in similar accuracy include the one of *Swamee* and *Jain* (1976):

$$\frac{1}{\sqrt{\lambda}} = -2\log\left[\frac{5.74}{\mathrm{Re}^{0.9}} + \frac{k}{3.7D}\right] \tag{3.21}$$

as well as the approximation of *Haaland* (1983):

$$\frac{1}{\sqrt{\lambda}} = -1.8\log\left[\frac{6.9}{\mathrm{Re}} + \left(\frac{k}{3.7D}\right)^{1.11}\right] \tag{3.22}$$

The *Reynolds number* describes the flow regime. It can be calculated as: *Reynolds number*

$$\mathrm{Re} = \frac{vD}{v} \tag{3.23}$$

where v (m²/s) stands for the *kinematic viscosity*. This parameter depends on the   *Kinematic*
water temperature and can be determined from the following equation: *viscosity*

$$v = \frac{497 \times 10^{-6}}{(T + 42.5)^{1.5}} \tag{3.24}$$

for T expressed in °C.
  The flow is:

1) laminar, if $\mathrm{Re} < 2000$,
2) critical (in transition), for $\mathrm{Re} \approx 2000 - 4000$, or
3) turbulent, if $\mathrm{Re} > 4000$.

The turbulent flows are predominant in distribution networks under normal operation. For example, within a typical range for the following parameters: v = 0.5-1.5 m/s, D = 50-1500 mm and T = 10-20 °C, the Reynolds number calculated by using equations 3.23 and 3.24 has a value of between 19,000 and 2,250,000.
  If for any reason $\mathrm{Re} < 4000$, equations 3.19-3.22 are no longer valid. The friction factor for the laminar flow conditions is then calculated as:

$$\lambda = \frac{64}{\mathrm{Re}} \tag{3.25}$$

As this usually results from very low velocities, this flow regime is not favourable in any way.

Once Re, k and D are known, the λ-factor can also be determined from the *The Moody* Moody diagram, shown in Figure 3.9. This diagram is in essence a graphic pre- *diagram* sentation of the Colebrook-White Equation.

In the turbulent flow regime, the Moody diagram shows a family of curves for different k/D ratios. This zone is split in two by the dashed line.

Transitional The first sub-zone is called the *transitional turbulence zone*, where the effect of *turbulence zone* the pipe roughness on the friction factor is limited compared to the impact of the Reynolds number (i.e. the viscosity).

Rough turbulence The curves in the second sub-zone of the *rough (developed) turbulence* are *zone* almost parallel, which clearly indicates the opposite situation where the Reynolds number has little influence on the friction factor. As a result, in this zone the Colebrook-White Equation can be simplified:

$$\frac{1}{\sqrt{\lambda}} = -2\log\left[\frac{k}{3.7D}\right] \tag{3.26}$$

For typical values of v, k, D and T, the flow rate in distribution pipes often drops within the rough turbulence zone.

Absolute The *absolute roughness* is dependent upon the pipe material and age. The *roughness* most commonly used values for pipes in good condition are given in Table 3.1. For new pipes, the absolute roughness specification will be given by the pipe manufacturer.

Examples of the inner surface for some of the materials in mostly new/smooth condition are to be seen in Figure 3.10. With the impact of corrosion, the k values can increase substantially. In extreme cases, severe corrosion will be taken into consideration by reducing the inner diameter; the corresponding effect on the calculation of friction losses will be the same.

The Hazen-Williams Equation

The Hazen-Williams Equation is an empirical equation widely used in practice. It is especially applicable for smooth pipes of medium and large diameters and pipes that are not attacked by corrosion (Bhave, 1991). The values of the Hazen-Williams constant, C_{hw} (-), for selected pipe materials and diameters are shown in Table 3.2.

Bhave states that the values in Table 3.2 are experimentally determined for a flow velocity of 0.9 m/s. A correction for the percentage given in Table 3.3 is therefore suggested if the actual velocity differs significantly. For example, the value of $C_{hw} = 120$ increases twice by 3% if the expected velocity is approximately a quarter of the reference value i.e. $C_{hw} = 127$ for v of, say, 0.22 m/s. On the other hand, for doubled velocity v = 1.8 m/s, $C_{hw} = 116$ i.e. 3% less than the original value of 120. However, such corrections do not significantly influence the friction loss calculation, and are, except for extreme cases, rarely applied in practice.

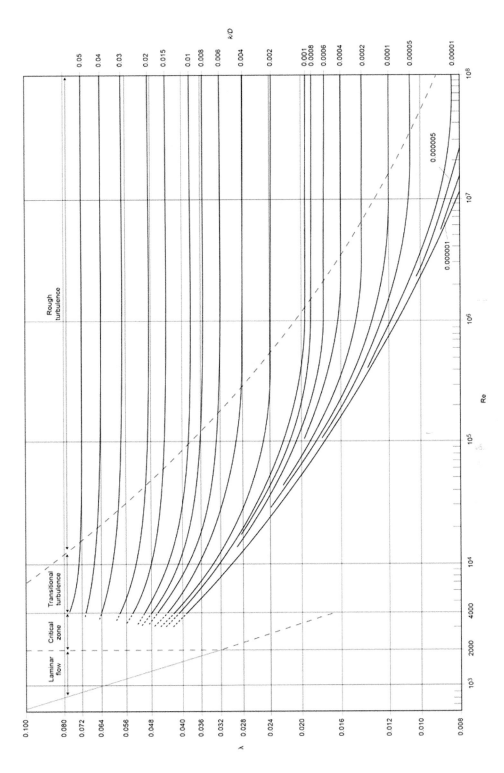

Figure 3.9 The Moody diagram

Table 3.1 Absolute roughness

Pipe material	k (mm)
Asbestos cement	0.015 – 0.03
Galvanised/coated cast iron	0.06 – 0.3
Uncoated cast iron	0.15 – 0.6
Ductile iron	0.03 – 0.06
Uncoated steel	0.015 – 0.06
Coated steel	0.03 – 0.15
Concrete	0.06 – 1.5
Plastic, PVC, polyethylene (PE)	0.02 – 0.05
Glass fibre	0.06
Brass, copper	0.003

Source: Wessex Water PLC, 1993.

Asbestos cement

Cement-lined ductile iron

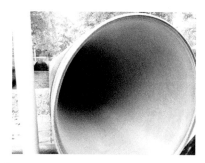

Cement-lined steel

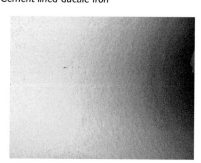

Concrete

Glass-reinforced plastic

Polyethilene

Figure 3.10 Examples of the inner surface of different pipe materials

Table 3.2 The Hazen-Williams factors

Pipe material	C_{hw} 75mm	C_{hw} 150mm	C_{hw} 300mm	C_{hw} 600mm	C_{hw} 1200mm
Uncoated cast iron	121	125	130	132	134
Coated cast iron	129	133	138	140	141
Uncoated steel	142	145	147	150	150
Coated steel	137	142	145	148	148
Galvanised iron	129	133	-	-	-
Uncoated asbestos cement	142	145	147	150	-
Coated asbestos cement	147	149	150	152	-
Concrete, minimum/maximum values	69/129	79/133	84/138	90/140	95/141
Pre-stressed concrete	-	-	147	150	150
PVC, brass, copper, lead	147	149	150	152	153
Wavy PVC	142	145	147	150	150
Bitumen/cement lined	147	149	150	152	153

Source: Bhave, 1991.

Table 3.3 Correction of the Hazen-Williams factors

C_{hw}	$v < 0.9m/s$ per halving	$v > 0.9m/s$ per doubling
less than 100	+5%	-5%
100-130	+3%	-3%
130-140	+1%	-1%
greater than 140	-1%	+1%

Source: Bhave, 1991.

Bhave also states that the Hazen-Williams Equation becomes less accurate for C_{hw}-values significantly below 100.

Table 3.4 The Manning factors

Pipe material	N $(m^{-1/3}s)$
PVC, brass, lead, copper, glass fibre	0.008 – 0.011
Pre-stressed concrete	0.009 – 0.012
Concrete	0.010 – 0.017
Welded steel	0.012 – 0.013
Coated cast iron	0.012 – 0.014
Uncoated cast iron	0.013 – 0.015
Galvanised iron	0.015 – 0.017

Source: Bhave, 1991.

The Manning Equation

The Manning Equation is another empirical equation used for the calculation of friction losses. In a slightly modified format, it also occurs in some literature under the name of *Strickler* (or *Gauckler*). The usual range of the N values ($m^{-1/3}s$) for typical pipe materials is given in Table 3.4.

Strickler (Gauckler) Equation

The Manning Equation is more suitable for rough pipes where N is greater than 0.015 m⁻¹/³s. It is frequently used for open channel flows rather than pressurised flows.

Comparison of the friction loss equations

The straightforward calculation of pipe resistance, being the main advantage of the Hazen-Williams and Manning equations, has lost its relevance as a result of developments in computer technology. The research also shows some limitations in the application of these equations compared to the Darcy-Weisbach Equation (Liou, 1998). Nevertheless, this is not necessarily a problem for engineering practice and the Hazen-Williams Equation in particular is still widely used in some parts of the world.

Figures 3.11 and 3.12 show the friction loss diagrams for a range of diameters and two roughness values calculated by each of the three equations. The flow in two pipes of different length, L = 200 and 2000 m respectively, is determined for velocity v = 1 m/s. Thus in all cases, for D in m and Q in m³/s:

$$Q = v\frac{D^2\pi}{4} = 0.7854D^2$$

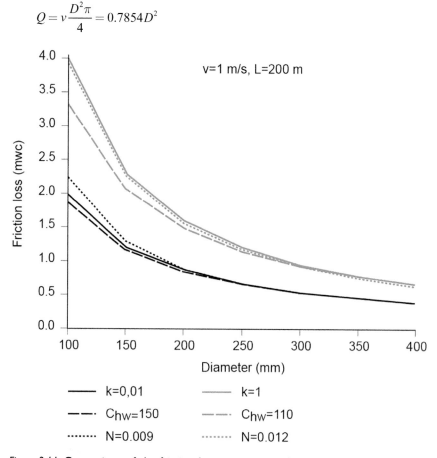

Figure 3.11 Comparison of the friction loss equations: mid-range diameters, v = 1 m/s, L = 200 m

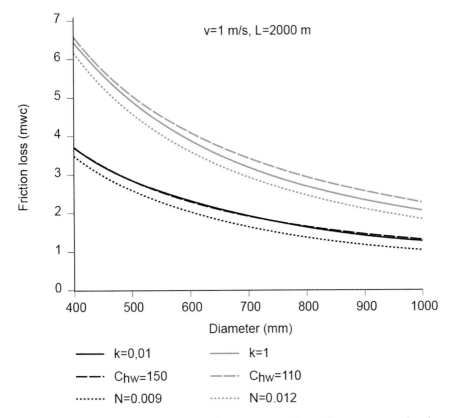

Figure 3.12 Comparison of the friction loss equations: large diameters, v = 1 m/s, L = 2000 m

The example shows little difference between the results obtained by the three different equations. However, the same roughness parameters have a different impact on the friction loss in the case of larger and longer pipes.

The difference in the results becomes larger if the roughness values are not properly chosen. Figure 3.13 shows the friction loss calculated using the roughness values suggested for PVC in tables 3.1, 3.2 and 3.4.

Hence, *the choice of a correct roughness value is more relevant than the choice of the friction loss equation itself.* Which of the values fits best to the particular case can be confirmed only by field measurements. In general, the friction loss will increase when there is:

1) an increase in pipe discharge,
2) an increase in pipe roughness,
3) an increase in pipe length,
4) a reduction in pipe diameter, or
5) a decrease in water temperature.

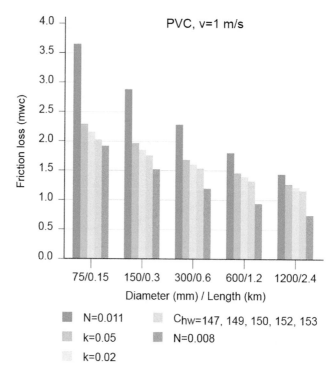

Figure 3.13 Comparison of the friction loss equations for various PVC roughness factors

In reality, the situations causing this to happen are:

- higher rates of consumption or leakage,
- corrosion growth, or
- network expansion.

The friction loss equations clearly point to the pipe diameter as the most sensitive parameter. The Darcy-Weisbach Equation shows that each halving of D (for example, from 200 to 100 mm) increases the head loss $2^5 = 32$ times! Moreover, the discharge variation will have a quadratic impact on the head losses, while these grow linearly with the increase of the pipe length. The friction losses are less sensitive to the change of the roughness factor, particularly in smooth pipes (an example is shown in Table 3.5). Lastly, the impact of water temperature variation on the head losses is marginal.

Table 3.5 Hydraulic gradient in pipe D = 300 mm, Q = 80 l/s, T = 10° C

Parameter	k = 0.01 mm	k = 0.1 mm	k = 1 mm	k = 5 mm
S (m/km)	3.3	3.8	6.0	9.9
Increase (%)	-	15	58	65

Problem 3-5

For pipe L = 450 m, D = 300 mm and flow rate of 120 l/s, calculate the friction loss by comparing the Darcy-Weisbach (k = 0.2 mm), Hazen-Williams (C_{hw} = 125) and Manning equations (N = 0.01). The water temperature can be assumed at 10 °C.

If demand grows at the exponential rate of 1.8 % annually, what will the friction loss be in the same pipe after 15 years? The assumed value of an increased absolute roughness in this period equals k = 0.5 mm.

Answer:

For a flow Q = 120 l/s and a diameter of 300 mm, the velocity in the pipe:

$$v = \frac{4Q}{D^2 \pi} = \frac{4 \times 0.12}{0.3^2 \times 3.14} = 1.70 \, m/s$$

Based on the water temperature, the kinematic viscosity can be calculated from Equation 3.24:

$$\upsilon = \frac{497 \times 10^{-6}}{(T + 42.5)^{1.5}} = \frac{497 \times 10^{-6}}{(10 + 42.5)^{1.5}} = 1.31 \times 10^{-6} \, m^2/s$$

The Reynolds number then becomes:

$$Re = \frac{vD}{\upsilon} = \frac{1.70 \times 0.3}{1.31 \times 10^{-6}} = 3.9 \times 10^5$$

For the value of relative roughness k/D = 0.2/300 = 0.00067 and the calculated Reynolds number, the friction factor λ can be determined from the Moody diagram in Figure 3.9 (λ ≈ 0.019). Based on the value of the Reynolds number (>> 4000), the flow regime is obviously turbulent and the same result can also be obtained by applying the Barr approximation. From Equation 3.20:

$$\lambda = \frac{0.25}{\log^2\left[\frac{5.1286}{Re^{0.89}} + \frac{k}{3.7D}\right]} = \frac{0.25}{\log^2\left[\frac{5.1286}{(3.9 \times 10^5)^{0.89}} + \frac{0.2}{3.7 \times 300}\right]} = 0.019$$

Finally, the friction loss from the Darcy-Weisbach Equation is determined as:

$$h_f = \frac{8\lambda L}{\pi^2 g D^5} Q^2 = \frac{\lambda L}{12.1 D^5} Q^2 = \frac{0.019 \times 450}{12.1 \times 0.3^5} 0.12^2 = 4.18 \, mwc$$

Applying C_{hw} = 125 to the Hazen-Williams Equation, the friction loss becomes:

$$h_f = \frac{10.68 L}{C_{hw}^{1.852} D^{4.87}} Q^{1.852} = \frac{10.68 \times 450}{125^{1.852} 0.3^{4.87}} 0.12^{1.852} = 4.37 \, mwc$$

Introducing a correction for the C_{hw} value of 3 %, as suggested in Table 3.3 based on the velocity of 1.7 m/s (almost twice the value of 0.9 m/s), yields a value of C_{hw}, which is reduced to 121. Using the same formula, the friction loss then becomes $h_f = 4.64$ mwc, which is 6 % higher than the initial figure.

Finally, applying the Manning Equation with the friction factor $N = 0.01$:

$$h_f = \frac{10.29 N^2 L}{D^{16/3}} Q^2 = \frac{10.29 \times 0.01^2 \times 450}{0.3^{16/3}} 0.120^2 = 4.10 \; mwc$$

With the annual growth rate of 1.8 %, the demand after 15 years becomes:

$$Q_{15} = 120 \left(1 + \frac{1.8}{100}\right)^{15} = 156.82 \; l/s$$

which, with the increase of the k value to 0.5 mm, yields the friction loss of 8.60 mwc by applying the Darcy-Weisbach Equation in the same way as shown above. The interim calculations give the following values of the parameters involved: $v = 2.22$ m/s, Re $= 5.1 \times 10^5$ and $\lambda = 0.023$. The final result represents an increase of more than 100 % compared to the original value of the friction loss (at the demand increase of approximately 30 %).

Self-study: Workshop problems 1.2.1 to 1.2.3 (Appendix 1)
        Spreadsheet Lessons 1-2 and 1-3 (Appendix 7)

3.2.2   Minor losses

Minor (in various literature *local* or *turbulence*) losses are usually caused by installed valves, bends, elbows, reducers, etc. Although the effect of the disturbance is spread over a short distance, for the sake of simplicity the minor losses are attributed to a cross section of the pipe. As a result, an instant drop in the hydraulic grade line will be registered at the place of obstruction (see Figure 3.14).

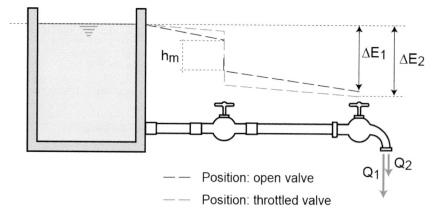

Figure 3.14 Minor loss caused by valve operation

Factors R_m and n_m from Equation 3.15 are uniformly expressed as:

$$R_m = \frac{8\xi}{\pi^2 g D^4} = \frac{\xi}{12.1 D^4}; \; n_m = 2 \qquad (3.27)$$

where ξ represents the *minor (local) loss coefficient*. This factor is usually deter-   *Minor loss*
mined by experiments. The values for most typical appendages are given in   *coefficient*
Appendix 5. A very detailed overview can be found in Idel'cik (1986).

The minor loss factors for various types of valves are normally supplied together
with the device. The corresponding equation may vary slightly from Equation
3.27, mostly in order to enable a diagram that is convenient for easy reading of the
values. In the example shown in Figure 3.15, the minor loss of a butterfly valve is

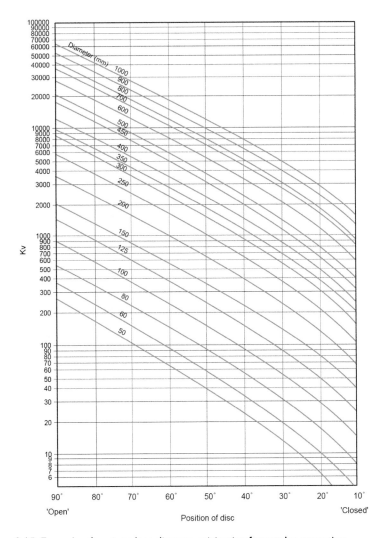

Figure 3.15 Example of a minor loss diagram originating from valve operation

calculated in mwc as: $h_m = 10Q^2/K_v^2$, for Q in m³/h. The K_v values can be determined from the diagram for different valve diameters and settings.

Substantial minor losses are measured in the following cases:

1) when the flow velocity is high, and/or
2) when there is a significant valve throttling in the system.

Such conditions commonly occur in pumping stations and in pipes with larger capacities where installed valves are regularly operated; given the magnitude of the head loss, the term 'minor' loss may not be appropriate in these situations.

The minor losses appear at numerous locations within a distribution network but are still comparatively smaller than the friction losses. In absence of accurate measurements, the minor-loss impact on overall head loss is typically represented through an adjustment of the roughness values (increased k and N or reduced C_{hw}). In such cases, $\Delta H \approx h_f$ is an acceptable approximation and the hydraulic gradient then becomes:

$$S = \frac{\Delta H}{L} \approx \frac{h_f}{L}$$ (3.28)

Equivalent pipe lengths

The other possibility for considering minor losses is to introduce what are referred to as *equivalent pipe lengths*. This approach is sometimes used in the design of indoor installations where the minor loss impact is simulated by assuming an increased pipe length (for example, up to 30-40 %) from the most critical end point.

3.3   Single pipe calculation

Summarised from the previous paragraphs, the basic parameters involved in the head-loss calculation of a single pipe using the Darcy-Weisbach Equation are:

1) length L,
2) diameter D,
3) absolute roughness k,
4) discharge Q,
5) piezometric head difference ΔH (i.e. the head loss), and
6) water temperature T.

The parameters derived from the above are:

7) velocity v, as a function (f) of Q and D,
8) hydraulic gradient, $S = f(\Delta H, L)$,
9) kinematic viscosity, $\upsilon = f(T)$,
10) Reynolds number, $Re = f(v, D, \upsilon)$, and
11) friction factor, $\lambda = f(k, D, Re)$

In practice, three of the six basic parameters are always included as an input:

- L, influenced by the consumers' location,
- k, influenced by the pipe material and its overall condition, and
- T, influenced by the ambient temperature.

The other three, D, Q and ΔH, are parameters with a major impact on pressures and flows in the system. Any of these parameters can be considered as the overall output of the calculation after setting the other two in addition to the three initial input parameters. The result obtained using this method answers one of the three typical questions that appear in practice:

1) What is the available head loss ΔH (and consequently the pressure) in a pipe of diameter D, when it conveys flow Q?
2) What is the flow Q that a pipe of diameter D can deliver if a particular maximum head loss ΔH_{max} (i.e. the minimum pressure p_{min}) is to be maintained?
3) What is the optimal diameter D of a pipe that has to deliver the required flow Q at a particular maximum head loss ΔH_{max} (i.e. minimum pressure p_{min})?

The calculation procedure in each of these cases is explained below. The form of the Darcy-Weisbach Equation linked to kinetic energy is more suitable in this case:

$$\Delta H \approx h_f = \frac{\lambda L}{12.1 D^5} Q^2 = \lambda \frac{L}{D} \frac{v^2}{2g}; \quad S = \frac{\lambda}{D} \frac{v^2}{2g} \qquad (3.29)$$

3.3.1   Pipe pressure

The input data in this type of the problem are: L, D, k, Q or v, and T, which yield ΔH (or S) as the result. The following procedure is to be applied:

1) For given Q and D, find the velocity, $v = 4Q/(D^2\pi)$.
2) Calculate Re from Equation 3.23.
3) Based on the Re value, choose the appropriate friction factor equation, 3.19-3.22 or 3.25, and determine the λ-factor. Alternatively, use the Moody diagram for an appropriate k/D ratio.
4) Determine ΔH (or S) using Equation 3.29.

The sample calculation has already been demonstrated in Problem 3-5.
   To be able to define the pressure head, $p/\rho g$, an additional input is necessary:

- the pipe elevation heads, Z, and
- the known (fixed) piezometric head, H, at one side.

There are two possible final outputs for the calculation:

1) If the downstream (discharge) piezometric head is specified, suggesting the minimum pressure to be maintained, the final result will show the required head/pressure at the upstream side i.e. at the supply point.

2)   If the upstream (supply) piezometric head is specified, the final result will show the available head/pressure at the downstream side i.e. at the discharge point.

Problem 3-6

The distribution area is supplied through a transportation pipe L = 750 m, D = 400 mm and k = 0.3 mm, with an average flow rate of 1260 m³/h. For this flow, the water pressure at the end of the pipe has to be maintained at a minimum of 30 mwc. What will the required piezometric level and also the pressure on the upstream side be in this situation? The average pipe elevation varies from Z_2 = 51 msl at the downstream side to Z_1 = 75 msl at the upstream side. The water temperature can be assumed at 10 °C.

Answer:

For flow Q = 1260 m³/h = 350 l/s and the diameter of 400 mm:

$$v = \frac{4Q}{D^2 \pi} = \frac{4 \times 0.35}{0.4^2 \times 3.14} = 2.79 \, m/s$$

For temperature T = 10 °C, the kinematic viscosity from Equation 3.24, υ = 1.31×10⁻⁶ m²/s. The Reynolds number then takes the value of:

$$Re = \frac{vD}{\upsilon} = \frac{2.79 \times 0.4}{1.31 \times 10^{-6}} = 8.5 \times 10^5$$

and the friction factor λ from Barr's Equation equals:

$$\lambda = \frac{0.25}{\log^2 \left[\frac{5.1286}{Re^{0.89}} + \frac{k}{3.7D} \right]} = \frac{0.25}{\log^2 \left[\frac{5.1286}{(8.5 \times 10^5)^{0.89}} + \frac{0.3}{3.7 \times 400} \right]} \approx 0.019$$

The friction loss from the Darcy-Weisbach Equation can be determined as:

$$h_f = \frac{\lambda L}{12.1 D^5} Q^2 = \frac{0.019 \times 750}{12.1 \times 0.4^5} 0.35^2 \approx 14 \, mwc$$

The downstream pipe elevation is given at Z_2 = 51 msl. By adding the minimum required pressure of 30 mwc to it, the downstream piezometric head becomes H_2 = 51 + 30 = 81 msl. On the upstream side, the piezometric head must be higher for the value of calculated friction loss, which produces a head of H_1 = 81 + 14 = 95 msl. Finally, the pressure on the upstream side will be obtained by deducting the upstream pipe elevation from this head. Hence $p_1/\rho g$ = 95 – 75 = 20 mwc. Due to configuration of the terrain in this example, the upstream pressure is lower than the downstream one. For the calculated friction loss, the hydraulic gradient S = h_f / L = 14 / 750 ≈ 0.019.

3.3.2   Maximum pipe capacity

For determination of the maximum pipe capacity, the input data are: L, D, k, ΔH (or S), and T. The result is flow Q.

Due to the fact that the λ factor depends on the Reynolds number i.e. the flow velocity, which is not known in advance, an iterative procedure is required here. The following steps have to be executed:

1) Assume the initial velocity (usually, v = 1 m/s).
2) Calculate Re from Equation 3.23.
3) Based on the Re value, choose the appropriate friction factor equation, 3.19-3.22 or 3.25, and calculate the λ value. For the selected Re and k/D values, the Moody diagram can also be used as an alternative.
4) Calculate the velocity after re-writing Equation 3.29:

$$v = \sqrt{\frac{2gDS}{\lambda}} \qquad (3.30)$$

If the values of the assumed and determined velocity differ substantially, steps 2-4 should be repeated by taking the calculated velocity as the new input. When a sufficient accuracy has been reached, usually after 2-3 iterations for flows in the transitional turbulence zone, the procedure is completed and the flow can be calculated from the final velocity. If the flow is in the rough turbulence zone, the velocity obtained in the first iteration will already be the final velocity, as the calculated friction factor will remain constant (being independent of the value of the Reynolds number).

If the Moody diagram is used, an alternative approach can be applied to determine the friction factor. The calculation starts by assuming the rough turbulence regime:

1) Read the initial λ value from Figure 3.9 based on the k/D ratio (or calculate it by applying Equation 3.26).
2) Calculate the velocity by applying Equation 3.30.
3) Calculate Re from Equation 3.23.

Check on the graph if the obtained Reynolds number corresponds to the assumed λ and k/D. If not, read the new λ value for the calculated Reynolds number and repeat steps 2 and 3. Once a sufficiently accurate λ value has been reached, the velocity calculated from this value will be the final velocity.

Both approaches are valid for a wide range of input parameters. The first approach is numerical, i.e. suitable for computer programming. The second is simpler for manual calculations; it is shorter and avoids estimation of the velocity in the first iteration. However, the second approach relies very much on accurate reading of the values from the Moody diagram.

Problem 3-7

For the system from Problem 3-6, calculate the maximum capacity that can be conveyed if the pipe diameter is increased to D = 500 mm and the head loss has been limited to 10 m per km of the pipe length. The roughness factor for the new pipe diameter can be assumed at k = 0.1 mm.

Answer:

Assume velocity $v = 1$ m/s. For the temperature $T = 10$ °C, the kinematic viscosity from Equation 3.24, $\upsilon = 1.31 \times 10^{-6}$ m²/s. With diameter $D = 500$ mm, the Reynolds number takes the value of:

$$Re = \frac{vD}{\upsilon} = \frac{1 \times 0.5}{1.31 \times 10^{-6}} = 3.8 \times 10^5$$

and the friction factor λ from Barr's Equation equals:

$$\lambda = \frac{0.25}{\log^2 \left[\dfrac{5.1286}{Re^{0.89}} + \dfrac{k}{3.7D} \right]} = \frac{0.25}{\log^2 \left[\dfrac{5.1286}{(3.8 \times 10^5)^{0.89}} + \dfrac{0.1}{3.7 \times 500} \right]} \approx 0.016$$

The new value of the velocity based on the maximum-allowed hydraulic gradient $S_{max} = 10 / 1000 = 0.01$ is calculated from Equation 3.30:

$$v = \sqrt{\frac{2gDS}{\lambda}} = \sqrt{\frac{2 \times 9.81 \times 0.5 \times 0.01}{0.016}} = 2.48 \ m/s$$

The result differs substantially from the assumed velocity and the calculation should be repeated in the second iteration with this value as a new assumption. Hence:

$$Re = \frac{2.48 \times 0.5}{1.31 \times 10^{-6}} = 9.5 \times 10^5$$

and the friction factor λ equals:

$$\lambda = \frac{0.25}{\log^2 \left[\dfrac{5.1286}{(9.5 \times 10^5)^{0.89}} + \dfrac{0.1}{3.7 \times 500} \right]} \approx 0.015$$

The new resulting velocity will be:

$$v = \sqrt{\frac{2 \times 9.81 \times 0.5 \times 0.01}{0.015}} = 2.57 \ m/s$$

which can be considered as a sufficiently accurate result, as any additional iteration that can be carried out is not going to change this value. Finally, the maximum flow that can be discharged at $S = 0.01$ equals:

$$Q = v \frac{D^2 \pi}{4} = 2.57 \frac{0.5^2 \times 3.14}{4} \approx 0.5 \, m^3/s \approx 1800 \, m^3/h$$

In the alternative approach, the initial λ value assumes that the rough turbulent zone can be read from the Moody diagram in Figure 3.9. For a value of $k/D = 0.1 / 500 = 0.0002$, it is approximately 0.014. The calculation from the rewritten Equation 3.26 gives:

$$\lambda = \frac{0.25}{\log^2\left[\dfrac{k}{3.7D}\right]} = \frac{0.25}{\log^2\left[\dfrac{0.1}{3.7\times500}\right]} = 0.0137$$

With this value:

$$v = \sqrt{\frac{2gDS}{\lambda}} = \sqrt{\frac{2\times9.81\times0.5\times0.01}{0.0137}} = 2.66\ m/s$$

and the Reynolds number then becomes:

$$Re = \frac{2.66\times0.5}{1.31\times10^{-6}} = 1.0\times10^6$$

which means that the new reading for λ is closer to the value of 0.015 (k/D = 0.0002). Repeated calculation of the velocity and the Reynolds number with this figure leads to a final result as in the first approach.

Self-study: Workshop problems 1.2.4 to 1.2.7 (Appendix 1)
Spreadsheet Lesson 1-4 (Appendix 7)

3.3.3   Optimal diameter

In the calculation of optimal diameters, the input data are: L, k, Q, ΔH (or S), and T. The result is diameter D.

The iteration procedure is similar to the one in the previous case, with the additional step of calculating the input diameter based on the assumed velocity:

1) Assume the initial velocity (usually, v = 1 m/s).
2) Calculate the diameter from the velocity/flow relation. $D^2 = 4Q/(v\pi)$.
3) Calculate Re from Equation 3.23.
4) Based on the Re value, choose the appropriate friction factor equation, 3.19-3.22 or 3.25, and determine the λ value. For the selected Re and k/D values, the Moody diagram can also be used instead.
5) Calculate the velocity from Equation 3.30.

If the values of the assumed and determined velocity differ substantially, steps 2-5 should be repeated by taking the calculated velocity as the new input.

After a sufficient accuracy has been achieved, the calculated diameter can be rounded up to a first higher (manufactured) size.

This procedure normally requires more iterations than for the calculation of the maximum pipe capacity. The calculation of the diameter from an assumed velocity is needed because the proper diameter assumption is often difficult and an inaccurate guess of D accumulates more errors than in the case of the assumption of velocity. For these reasons, the second approach in Section 3.3.2 is not recommended here.

Problem 3-8

If the flow from the previous problem has to be doubled to Q = 3600 m³/h, calculate the diameter that would be sufficient to convey it without increasing the hydraulic gradient. The other input parameters remain the same as in Problem 3-7.

Answer:
Assume velocity v = 1 m/s. Based on this velocity, the diameter D:

$$D = 2\sqrt{\frac{Q}{v\pi}} = 2 \times \sqrt{\frac{1}{1 \times 3.14}} = 1.128\,m$$

and the Reynolds number:

$$Re = \frac{vD}{\upsilon} = \frac{1 \times 1.128}{1.31 \times 10^{-6}} = 8.6 \times 10^{5}$$

The friction factor λ from Barr's Equation equals:

$$\lambda = \frac{0.25}{\log^2\left[\dfrac{5.1286}{Re^{0.89}} + \dfrac{k}{3.7D}\right]} = \frac{0.25}{\log^2\left[\dfrac{5.1286}{(8.6 \times 10^5)^{0.89}} + \dfrac{0.1}{3.7 \times 1128}\right]} \approx 0.0135$$

and at S_{max} = 0.01 the velocity from Equation 3.30 becomes:

$$v = \sqrt{\frac{2gDS}{\lambda}} = \sqrt{\frac{2 \times 9.81 \times 1.128 \times 0.01}{0.0135}} = 4.04\,m/s$$

The result is substantially different than the assumed velocity and the calculation has to be continued with several more iterations. The results after iterating the same procedure are shown in the following table:

Iter.	v_{ass} (m/s)	D (mm)	Re (-)	λ (-)	v_{calc} (m/s)
2	4.04	561	1.7×10⁶	0.0141	2.79
3	2.79	676	1.4×10⁶	0.0139	3.09
4	3.09	642	1.5×10⁶	0.0139	3.01
5	3.01	650	1.5×10⁶	0.0139	3.03

with the final value for the diameter of D = 650 mm. The manufactured size would be, say, D = 700 mm.

Self-study: Workshop problem 1.2.8 (Appendix 1)
Spreadsheet Lesson 1-5 (Appendix 7)

3.3.4 Pipe charts and tables

Straightforward determination of the required pressures, flows or diameters is possible by using the *pipe charts* or *pipe tables*. These are created by combining the Darcy-Weisbach and Colebrook-White equations. Substituting λ and Re in Equation 3.19, by using equations 3.30 and 3.23 respectively, yields the following equation:

$$v = -2\sqrt{2gDS}\ \log\left[\frac{2.51\upsilon}{D\sqrt{2gDS}} + \frac{k}{3.7D}\right] \qquad (3.31)$$

For a fixed k value and water temperature (i.e. the viscosity), the velocity is calculated for common ranges of D and S. The values for v, D, S and Q are then plotted or sorted in a tabular form (see Appendix 6).

The chart in Figure 3.16 shows an example of a flow rate of 20 l/s (the top axis) passing through a pipe of diameter D = 200 mm (the bottom axis). From the intersection of the lines connecting these two values it emerges that the corresponding velocity (the left-hand axis) and hydraulic gradient (the right-hand axis) would be approximately 0.6 m/s and 2 m/km, respectively. The same flow rate in a pipe

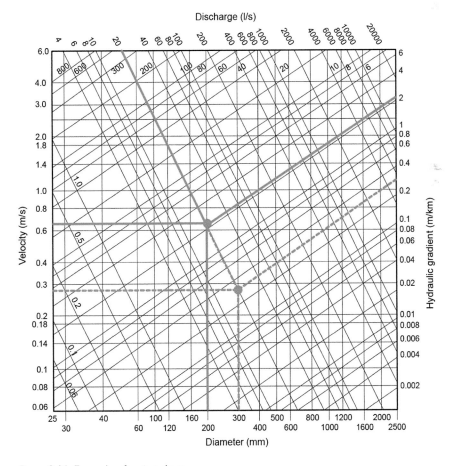

Figure 3.16 Example of a pipe chart

D = 300 mm yields much lower values: the velocity would be below 0.3 m/s and the gradient approximately 0.3 m/km.

It is important to note that the particular graph or table is valid for *one single* roughness value and *one single* water temperature. Although the variation of these parameters has a smaller effect on the friction loss than the variation of D, v or Q, this limits the application of the tables and graphs if the values specifically for k differ substantially from those used in the creation of the table/graph. As an example, Table 3.6 shows the difference in the calculation of hydraulic gradients for a range of values for k and T.

Table 3.6 Hydraulic gradient S (-) in pipe D = 400 mm at Q = 200 l/s

Parameter	k = 0.01 mm	k = 0.1 mm	k = 1 mm	k = 5 mm
T = 10 °C	0.0044	0.0052	0.0082	0.0133
T = 20 °C	0.0042	0.0051	0.0081	0.0132
T = 40 °C	0.0040	0.0049	0.0081	0.0132

In former times, pipe charts and tables were widely used for hydraulic calculations. Although they have become less relevant since the advent of PC spreadsheet programmes, they are nevertheless still a useful help in providing quick and straightforward estimates of pipe discharges for given design layouts.

Problem 3-9

Using the pipe tables in Appendix 6, determine the maximum discharge capacity for pipe D = 800 mm for the following roughness values: k = 0.01, 0.5, 1 and 5 mm and the maximum-allowed hydraulic gradients of S = 0.001, 0.005, 0.01 and 0.02, respectively. The water temperature can be assumed at T = 10 °C.

Answer:

The following table shows the results read for pipe D = 800 mm from the tables in Appendix 6:

Discharge flows Q (l/s) for pipe D = 800 mm (for T = 10 °C).

Parameter	k = 0.01 mm	k = 0.5 mm	k = 1 mm	k = 5 mm
S = 0.001	559.5	465.6	432.3	348.3
S = 0.005	1336.2	1052.5	972.8	780.0
S = 0.01	1936.8	1492.6	1377.8	1103.6
S = 0.02	2800.1	2115.1	1950.6	1561.1

The results suggest the following two conclusions:

1. For fixed values of S, the discharge capacity is reduced by the increase in the roughness value. In other words, the pipes start to lose their conveying capacity as they get older, which is reflected in reality by the drop in demand and/or pressure.

2.   The discharge at the fixed k value will increase by allowing the higher hydraulic gradient. In other words, if a greater friction loss is allowed in the network, more water will be distributed but at higher operational costs (because of additional pumping).

3.3.5   Equivalent diameters

During the planning for network extensions or renovations, the alternatives of laying a single pipe or pipes either connected in parallel or laying them in series is sometimes compared. To provide a hydraulically equivalent system, *the capacity and hydraulic loss along the considered section should remain unchanged* in all the options. These pipes are then of *equivalent diameter* (see Figure 3.17).

Each pipe in the parallel arrangement creates the same friction loss, which is equal to the total loss at the section. The total capacity is the sum of the flows in all the pipes. Hence, for *n* pipes it is possible to write:

$$\Delta H_{equ} = \Delta H_1 = \Delta H_2 = ... = \Delta H_n$$
$$Q_{equ} = Q_1 + Q_2 + ... + Q_n$$

Pipes in parallel are more frequently of the same diameter, allowing for easier maintenance and handling of irregular situations. Furthermore, they will often be laid in the same trench i.e. along the same route and can therefore be assumed to be of the same length, in which case the slope of the hydraulic grade line for all the pipes will be equal. Nevertheless, the equation $S_{equ} = S_1 = S_2 ..... = S_n$ is not always true as *pipes connected in parallel need not necessarily be of identical length.*

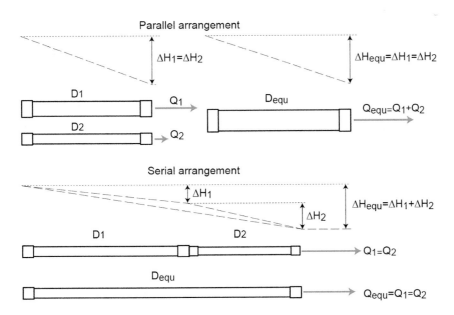

Figure 3.17. Equivalent diameters

For pipes in series, the basic hydraulic condition is that each pipe carries the same flow rate. The total energy loss is the sum of the losses in all the pipes. If written for n pipes:

$$\Delta H_{equ} = \Delta H_1 + \Delta H_2 + ... + \Delta H_n$$
$$Q_{equ} = Q_1 = Q_2 = ... = Q_n$$

Equation $S_{equ} = S_1 + S_2 + ... + S_n$, is never true. In the hypothetical case, $S_{equ} = S_1 = S_2 = ... = S_n$.

The hydraulic calculation of the equivalent diameters further proceeds based on the principles of the single pipe calculation, as explained in Section 3.3.

Problem 3-10

A pipe $L = 550$ m, $D = 400$ mm, and $k = 1$ mm transports a flow of 170 l/s. Due to an extension of the system this capacity needs to grow to 250 l/s. Two alternatives to solve this problem are considered:

1.   to lay a parallel pipe of smaller diameter on the same route, or
2.   to lay a parallel pipe of the same diameter on a separate route with a total length $L = 800$ m.

Using the hydraulic tables for water temperature $T = 10\ °C$:

   a)   determine the diameter of the pipe required to supply the surplus capacity of 80 l/s in the first alternative, and
   b)   determine the discharge of the second pipe $D = 400$ mm in the second alternative.

In both cases, the absolute roughness of the new pipes can be assumed to be $k = 0.1$ mm.

Answers:

In the hydraulic tables in Appendix 6 (for $T = 10\ °C$), the diameter $D = 400$ mm conveys the flow $Q = 156.6$ l/s for the hydraulic gradient $S = 0.005$ and $Q = 171.7$ l/s for $S = 0.006$. Assuming linear interpolation (which introduces negligible error), the flow of 170 l/s will be conveyed at $S = 0.0059$, leading to a friction loss $h_f = S \times L = 0.0059 \times 550 = 3.25$ mwc. This value is to be maintained in the design of the new parallel pipe.

Laying the second pipe in the same trench (i.e. with the same length) should provide an additional flow of 80 l/s. From the hydraulic tables for $k = 0.1$ mm the following closest discharge values can be read:

Discharge flows (l/s) for pipe k = 0.1 mm

Parameter	D = 250 mm	D = 300 mm
S = 0.005	57.2	92.5
S = 0.006	62.9	101.7

which suggests that the manufactured diameter of 300 mm is the final solution. The flow rate to be conveyed at S = 0.0059 is Q = 100.8 l/s (after interpolation) leading to a total supply of 270.8 l/s.

In the second case, the parallel pipe D = 400 mm follows an alternative route with a total length of L = 800 m. The value of the hydraulic gradient will be consequently reduced to S = 3.25 / 800 = 0.0041. The hydraulic tables give the following readings closest to this value:

Discharge flows (l/s) for pipe k = 0.1 mm

Parameter	D = 400 mm
S = 0.004	175.6
S = 0.005	197.3

Despite the longer route, this pipe is sufficiently large to convey capacities far beyond the required 80 l/s. For S = 0.0041, discharge Q = 177.8 l/s and the total supplying capacity from both pipes equals 347.8 l/s. Hence, more water but at higher investment costs.

Self-study: Workshop problems 1.2.9 to 1.2.11 (Appendix 1)
              Spreadsheet Lessons 2-1 to 2-5 (Appendix 7)

3.4   Serial and branched networks

The calculation of serial and branched networks is entirely based on the methods used for single pipes. The differences in hydraulic performance occur between the branched systems with one supply point and those that have more than one supply point.

3.4.1   Supply at one point

With known nodal demands, the flows in all pipes can easily be determined by applying the Continuity Equation (Equation 3.5), starting from the end points of the system (Figure 3.18).

   If the diameters of the pipes are also known, the head-loss calculation follows the procedure in Section 3.3.1, resulting in the hydraulic gradient S for each pipe. In the next step the piezometric heads, and consequently the pressures, will be calculated for each node starting from the node assumed to have the minimum pressure. In this respect potentially critical nodes are those with either a high elevation and/or nodes located far away from the source. Adding or subtracting the head losses for each pipe, depending on the flow direction, will determine all the other heads including the required piezometric head at the supply point. Calculation of the piezometric heads in the opposite direction, starting from the known value at the source, is also possible; this shows the pressures in the system available for a specified head at the supplying point.

   In situations where pipe diameters have to be designed, the maximum hydraulic gradient allowed must be included in the calculation input. The iterative procedure from Section 3.3.3 or the pipe charts/tables is required here, leading to

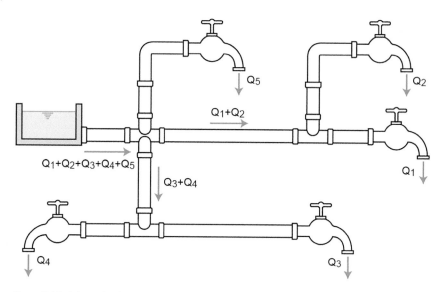

Figure 3.18 A branched network with a single supply point

actual values for the hydraulic gradient of each pipe based on the best available (manufactured) diameter. Finally, the pressures in the system will also be determined either by setting the minimum pressure criterion or the head available at the supply point. Throughout the latter process, some of the pipe diameters may need to be further increased in order to satisfy the minimum pressure required in the network.

3.4.2 Supply at several points

For more than one supply point, the contribution from each source may differ depending on its piezometric head and distribution of nodal demands in the system. In this case, flows in the pipes connecting the sources are not directly known from the Continuity Equation. These flows can change their rate and even reverse the direction due to a variation in nodal demands. Figures 3.19 and 3.20 show an example of anticipated demand increase in node 1 which reverses the flow direction in the pipe connecting the sources.

Except for the chosen source, fixed conditions are required for all the other sources existing in the system: a head, discharge or the hydraulic gradient of the connecting pipe(s). For the remainder, the calculation proceeds in precisely the same manner as in the case of one supply point. Alternatively, in the case of known source heads, an iterative calculation can be carried out after linking the sources into dummy loops by connecting them with dummy pipe(s). The general principles of the looped network calculation are explained in the following paragraph.

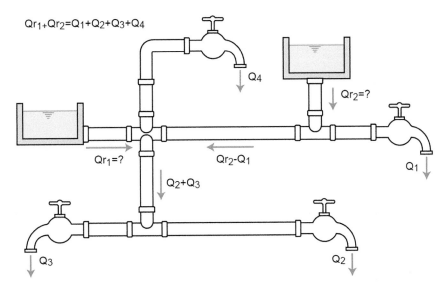

Figure 3.19 A branched network with two supply points

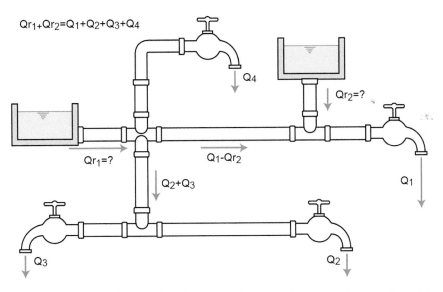

Figure 3.20 A branched network with two supply points, showing an increase in nodal flow Q_1

Problem 3-11

For the branched system shown below, calculate the pipe flows and nodal pressures for the surface level in the reservoir of H = 50 msl. Assume for all pipes k = 0.1 mm, and water temperature of 10° C.

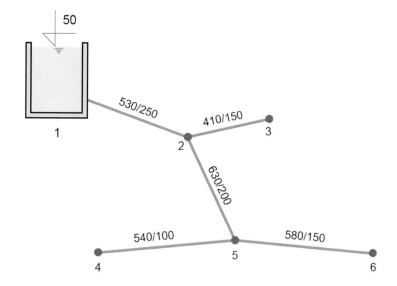

Nodes: ID
Pipes: L(m)/D(mm)

	1	2	3	4	5	6
Z (msl)	-	12	22	17	25	20
Q (l/s)	-75.6	10.4	22.1	10.2	18.5	14.4

Answer:

The total supply from the reservoir equals the sum of all the nodal demands, which is 75.6 l/s. Applying the Continuity Equation in each node (Equation 3.5), both the flow rate and its direction can be determined; each pipe conveys the flow that is the sum of all the downstream nodal demands. The pipe friction loss will be further calculated by the approach discussed in Problem 3-6 (Section 3.3.1). If the hydraulic tables from Appendix 6 are used, the friction loss will be calculated from the interpolated hydraulic gradients at a given diameter and flow rate (for fixed k and T). The results of the calculation applying the Darcy-Weisbach Equation are shown in the following table.

Pipe	D (mm)	Q (l/s)	v (m/s)	Re (-)	λ (-)	S (-)	L (m)	h_f (mwc)
1-2	250	75.6	1.54	2.9×10^5	0.018	0.0086	530	4.55
2-3	150	22.1	1.25	1.4×10^5	0.020	0.0108	410	4.43
2-5	200	43.1	1.37	2.1×10^5	0.190	0.0090	630	5.70
5-4	100	10.2	1.30	9.9×10^4	0.022	0.0192	540	10.38
5-6	150	14.4	0.81	9.4×10^4	0.021	0.0048	580	2.78

Finally, the pressure in each node is calculated by subtracting the friction losses starting from the reservoir surface level and further deducting the nodal elevation from the piezometric heads obtained in this way. The final results are shown in the following table and figure.

	1	2	3	4	5	6
H(msl)	50	45.45	41.02	29.37	39.75	36.96
p/ρg (mwc)	-	33.45	19.02	12.37	14.75	16.96

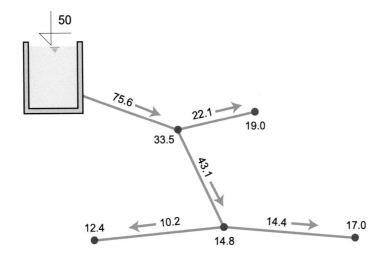

Nodes: p/ρg (mwc)
Pipes: Q(l/s)

The lowest pressure appears to be in node 4 (12.4 mwc) resulting from a relatively small diameter (causing large friction loss) of pipe 5-4.

Self-study: Workshop problems 1.3.1 to 1.3.5 (Appendix 1)
              Spreadsheet Lessons 3-1 to 3-2 (Appendix 7)

3.5   Looped networks

The principles of calculation as applied to single pipes are not sufficient in the case of looped networks. Instead, a system of equations is required which can be solved by numerical algorithm. This system of equations is based on the analogy of two electricity laws known in physics as *Kirchoff's Laws*. Translated into water distribution networks, these laws state that: *Kirchoff's Laws*

1) The sum of all the ingoing and outgoing flows in each node equals zero ($\Sigma Q_i = 0$).
2) The sum of all the head losses along the pipes that compose a complete loop equals zero ($\Sigma \Delta H_i = 0$).

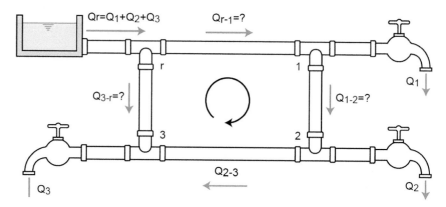

Figure 3.21 A looped network

The first law is essentially the Mass Conservation Law, resulting in the Continuity Equation which must be valid for each node in the system.

From the second law, it emerges that the hydraulic grade line along one loop is also continuous, just as the flow in any node is. The number of equations that can be formulated by applying this law equals therefore the number of loops. For example, in the simple network in the clockwise direction in Figure 3.21, this yields:

$$(H_r\text{-}H_1) + (H_1\text{-}H_2) + (H_2\text{-}H_3) - (H_r\text{-}H_3) = 0$$

3.5.1   Hardy Cross Methods

Two similar iterative procedures can be derived from Kirchoff's Laws:

1) The Balancing Head Method.
2) The Balancing Flow Method.

These methods, known in literature under the name of Hardy Cross (published in 1936 and developed further by Cornish in 1939), are used to calculate the pipe flows and nodal piezometric heads in looped systems for a given input, which is:

- for pipes: length L, diameter D, absolute roughness k and minor loss factor ξ,
- for nodes: nodal discharge Q and elevation Z.

The head in at least one node has to be fixed, which will influence the pressure in the rest of the system. This is usually a supply point.

An iterative calculation of the loops (nodes) is executed by following the following steps:

Balancing Head Method

1) Flows from an initial guess are assigned to each pipe. However, these must satisfy the Continuity Equation in all nodes.

2)   Head loss in each pipe is calculated starting from Equation 3.15.
3)   The sum of the head losses along each loop is checked.
4)   If the head-loss sum at any loop is outside of the required accuracy range, $0 \pm \varepsilon_{\Delta H}$ mwc, the following flow correction has to be introduced for each pipe within that loop (total n pipes):

$$
\delta Q_j = \frac{-\sum_{j=1}^{n} \Delta H_j}{2 \sum_{j=1}^{n} \left| \dfrac{\Delta H_j}{Q_j} \right|}
\tag{3.32}
$$

5)   Correction δQ is applied throughout the loop in a consistent direction: clockwise or anti-clockwise. This has implications for the value of the pipe flows, which will be negative if their direction counters the adopted orientation. The positive/negative sign of the correction should also be taken into account while adding it to the current pipe flow.
6)   The iteration procedure is carried out for the new flows, $Q + \delta Q$, repeated in steps 2-5, until $\varepsilon_{\Delta H}$ is satisfied for all the loops.
7)   After the iteration of flows and head losses is completed, the pressures in the nodes can be determined from the reference node(s) with fixed head(s), taking into account the flow directions.

The calculation proceeds simultaneously for all the loops in the network, with their corresponding corrections δQ being applied in the same iteration. In the case of pipes shared between two neighbouring loops, the sum of the two δQ corrections should be applied. The flow continuity in the nodes will not be affected in this case; assuming uniform orientation for both loops will reverse the sign of the composite δQ in one of them.

Unlike with branched network hydraulic calculations, multiple sources will not complicate the calculation provided that they are an integral part of the loops i.e. connected to the rest of the network with more than one pipe. If the system is supplied from one or more sources connected with single pipes, the number of unknowns increases; technically, the network is no longer looped but partly also branched i.e. with a combined configuration. Dummy loops have to be created in this case by connecting these 'branched' sources with dummy pipes with fictitious L, D and k, but with a fixed ΔH equal to the surface elevation difference between the connected reservoirs. This value has to be maintained throughout the entire iteration process.

Balancing Flow Method

1)   The estimated piezometric heads are initially assigned to each node in the system, except for the reference i.e. fixed pressure node(s). An arbitrary distribution is allowed in this case.
2)   The piezometric head difference is determined for each pipe.
3)   Flow in each pipe is determined starting from the equation for head loss, 3.15.

4) The continuity equation is checked in each node.
5) If the sum of flows in any node is out of the requested accuracy range, $0 \pm \varepsilon_Q$ m³/s, the following piezometric head correction has to be introduced in that node (n is the number of pipes connected in the node):

$$\delta H_i = \frac{2\sum_{i=1}^{n} Q_i}{\sum_{i=1}^{n} \left| \frac{Q_i}{\Delta H_i} \right|} \tag{3.33}$$

6) The iteration procedure is continued with the new heads, $H + \delta H$, repeated in steps 2-5, until ε_Q is satisfied for all the nodes.

A faster convergence is reached than in the Balancing Head Method by applying the corrections consecutively. As a consequence, the flow continuity in some of the nodes will include the pipe flows calculated from the piezometric heads of the surrounding nodes from the same iteration.

The required calculation time for both methods is influenced by the size of the network. The Balancing Head Method involves systems with a smaller number of equations, equal to the number of loops, which saves time while carrying out the calculation manually. The Balancing Flow Method requires a larger system of equations, equal to the number of nodes. However, this method excludes the identification of loops, which is of advantage for computer programming. The layout and operation of the system may have an impact on convergence in both methods. In general, faster convergence is reached by the Balancing Head Method.

The Hardy Cross Methods were widely used in the pre-computer era. The first hydraulic modelling software in water distribution was also based on these methods, with several improvements introduced in the meantime. The Balancing Flow Method was first developed for computer applications while the Balancing Head Method remained a faster approach for manual calculations of simple looped networks. Both methods are programmable in spreadsheet form, which helps in reducing the calculation time in such cases. However, the explanations and spreadsheet applications given in this book are only intended to enable a better understanding of the genesis of concepts used in the development of powerful network modelling software widely used in engineering practice nowadays.

Problem 3-12

To improve the conveyance of the system from Problem 3-11, nodes one and four as well as nodes three and six have been connected with pipes $D = 100$ mm and $L = 1200$ and 1040 m, respectively ($k = 0.1$ mm in both cases). Calculate the pipe flows and nodal pressures for such a system by applying the Balancing Head Method.

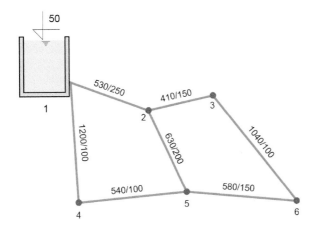

Nodes: ID
Pipes: L(m)/D(mm)

Answer:

Two loops are created from the branched system after adding the new pipes. The calculation starts by distributing the pipe flows arbitrarily, but satisfying the Continuity Equation in each node. The next step is to calculate the friction losses in each loop, as the following tables show (negative values mean the reverse direction, from the right node to the left node):

LOOP ONE – Iteration One

Pipe	D (mm)	Q (l/s)	v (m/s)	h_f (mwc)
1-2	250	50.6	1.03	2.12
2-5	200	20.2	0.64	1.36
5-4	100	−14.8	−1.88	−21.17
4-1	100	−25.0	−3.18	−129.89

The sum of all the friction losses, which should be equal to 0 for correct flow rate values, is in this case $\Sigma h_f = -147.59$ mwc (selecting the clockwise direction). Thus, all the pipe flows in Loop One must be corrected in the new iteration. From Equation 3.32 (for $\Delta H = h_f$) this correction becomes $\delta Q = 10.96$ l/s.
    In the case of Loop Two:

LOOP TWO – Iteration One

Pipe	D (mm)	Q (l/s)	v (m/s)	h_f (mwc)
5-2	200	−20.20	−0.64	−1.36
2-3	150	20.00	1.13	3.66
3-6	100	−2.10	−0.27	−1.06
6-5	150	−16.50	−0.93	−3.60

The sum of all the friction losses in this case is $\Sigma h_f = -2.35$ mwc, which is closer to the final result but also requires another flow correction. After applying Equation 3.32, $\delta Q = 1.21$ l/s.

In the second iteration, the following results are achieved after applying the pipe flows $Q + \delta Q$:

LOOP ONE – Iteration Two

Pipe	D (mm)	Q (l/s)	v (m/s)	h_f (mwc)
1-2	250	61.56	1.25	3.07
2-5	200	29.95	0.95	2.85
5-4	100	-3.84	-0.49	-1.66
4-1	100	-14.04	-1.79	-42.53

$\Sigma h_f = -38.27$ mwc and therefore $\delta Q = 5.31$ l/s

LOOP TWO – Iteration Two

Pipe	D (mm)	Q (l/s)	v (m/s)	h_f (mwc)
5-2	200	-29.95	-0.95	-2.85
2-3	150	21.21	1.20	4.10
3-6	100	-0.89	-0.11	-0.23
6-5	150	-15.29	-0.87	-3.12

$\Sigma h_f = -2.10$ mwc and therefore $\delta Q = 1.40$ l/s

The new flow in pipes 2-5 shared between the loops has been obtained by applying the correction δQ of both loops, i.e. 20.20 + 10.96 - 1.21 = 29.95 l/s. This pipe in Loop Two has a reversed order of nodes and therefore Q_{5-2} = -20.20 + 1.21 – 10.96 = -29.95 l/s. Hence, the corrected flow of the shared pipe maintains the same value in both loops, once with a positive and once with a negative sign.

In the remaining calculations:

LOOP ONE – Iteration Three

Pipe	D (mm)	Q (l/s)	v (m/s)	h_f (mwc)
1-2	250	66.86	1.36	3.60
2-5	200	33.85	1.08	3.60
5-4	100	1.46	0.19	0.29
4-1	100	-8.74	-1.11	-17.17

$\Sigma h_f = -9.69$ mwc and therefore $\delta Q = 2.09$ l/s

LOOP TWO – Iteration Three

Pipe	D (mm)	Q (l/s)	v (m/s)	h_f (mwc)
5-2	200	−33.85	−1.08	−3.60
2-3	150	22.61	1.28	4.63
3-6	100	0.51	0.06	0.09
6-5	150	−13.89	−0.79	−2.60

Σh_f = -1.48 mwc and therefore δQ = 1.11 l/s

LOOP ONE – Iteration Four

Pipe	D (mm)	Q (l/s)	v (m/s)	h_f (mwc)
1-2	250	68.95	1.40	3.82
2-5	200	34.83	1.11	3.80
5-4	100	3.55	0.45	1.43
4-1	100	−6.65	−0.85	−10.25

Σh_f = −1.20 mwc and therefore δQ = 0.29 l/s

LOOP TWO – Iteration Four

Pipe	D (mm)	Q (l/s)	v (m/s)	hf (mwc)
5-2	200	−34.83	−1.11	−3.80
2-3	150	23.72	1.34	5.07
3-6	100	1.62	0.21	0.66
6-5	150	−12.78	−0.72	−2.22

Σh_f = −0.29 mwc and therefore δQ = 0.16 l/s

LOOP ONE – Iteration Five

Pipe	D (mm)	Q (l/s)	v (m/s)	h_f (mwc)
1-2	250	69.24	1.41	3.85
2-5	200	34.96	1.11	3.82
5-4	100	3.84	0.49	1.65
4-1	100	−6.36	−0.81	−9.44

Σh_f = −0.12 mwc and therefore δQ = 0.03 l/s

LOOP TWO – Iteration Five

Pipe	D (mm)	Q (l/s)	v (m/s)	h_f (mwc)
5-2	200	-34.96	-1.11	-3.82
2-3	150	23.88	1.35	5.14
3-6	100	1.78	0.23	0.78
6-5	150	-12.62	-0.71	-2.17

Σh_f = -0.08 mwc and therefore δQ = 0.04 l/s

LOOP ONE – Iteration Six

Pipe	D (mm)	Q (l/s)	v (m/s)	h_f (mwc)
1-2	250	69.26	1.41	3.85
2-5	200	34.94	1.11	3.82
5-4	100	3.86	0.49	1.68
4-1	100	-6.34	-0.81	-9.36

Σh_f = -0.02 mwc and therefore δQ = 0.00 l/s

LOOP TWO – Iteration Six

Pipe	D (mm)	Q (l/s)	v (m/s)	h_f (mwc)
5-2	200	-34.94	-1.11	-3.82
2-3	150	23.92	1.35	5.15
3-6	100	1.82	0.23	0.82
6-5	150	-12.58	-0.71	-2.16

Σh_f = -0.01 mwc and therefore δQ = 0.01 l/s

As the tables show, this method already provides fast convergence after only the first two to three iterations and a continuation of the calculations does not add much to the accuracy of the results despite being time consuming, specifically in the case of manual calculations.

Determination of the nodal pressures will be carried out in the same way as in the case of branched systems: starting from the supply point with a fixed piezometric head, or from the point with minimum pressure required in the system. In each pipe, the unknown piezometric head on one side is obtained either by adding the pipe friction loss to the known downstream piezometric head, or by subtracting the pipe friction loss from the known upstream piezometric head. The final results are shown in the figure below.

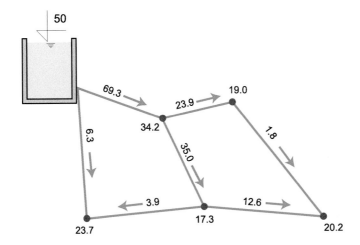

Nodes: p/ρg (mwc)
Pipes: Q(l/s)

Self-study: Workshop problems 1.4.1 to 1.4.3 (Appendix 1)
Spreadsheet Lessons 4-1 and 4-2 (Appendix 7)

3.5.2   The Linear Theory

The Newton Raphson and the Linear Theory methods succeeded the Hardy-Cross Methods as the main approach for solving non-linear network governing equations. The Linear Theory Method was developed by Wood and Charles (1972) and involves a remarkably simple linearization. When using the Darcy-Weisbach Equation that is written like this:

$$\Delta H = (R_f + R_m)|Q|Q = UQ \qquad (3.34)$$

where:

$$U = \frac{(\lambda L + \xi D)|Q|}{12.1 D^5} \qquad (3.35)$$

The absolute value of Q helps to distinguish between different flow directions (the +/- sign).

The following can be written for node i, assuming the inflow to the node has a negative sign (Figure 3.22):

$$Q_i - \sum_{j=1}^{n} Q_{ij} = 0 \qquad (3.36)$$

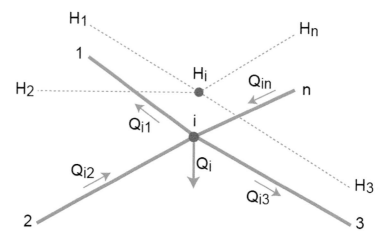

Figure 3.22 The Linear Theory application

Index n represents the total number of pipes/surrounding nodes connected to node i, while Q_i is the nodal demand (outflow).

Equation 3.36 is satisfied in the iteration procedure with specified accuracy ε_i, for each node. Thus:

$$Q_i - \sum_{j=1}^{n} Q_{ij} = \pm\varepsilon_i = f(H_i) \tag{3.37}$$

and after combining equations 3.34 and 3.36:

$$\pm\varepsilon_i = f(H_i) = Q_i - \sum_{j=1}^{n} \frac{H_j}{U_{ij}} + H_i \sum_{j=1}^{n} \frac{1}{U_{ij}} \tag{3.38}$$

The system of linear equations given in 3.38 equals the total number of nodes in the network. The solution of these linear equations can be achieved by any standard procedure (i.e. algorithms for inversion and decomposition). For example, one could apply the Newton-Raphson Method that would give the nodal head in the (k+1)ᵗʰ iteration as follows:

$$H_i^{(k+1)} = H_i^{(k)} - \frac{f(H_i^{(k)})}{f'(H_i^{(k)})} \tag{3.39}$$

Plugging 3.38 into 3.39 yields:

$$f(H_i^{(k)}) = Q_i - \sum_{j=1}^{n} \frac{H_j^{(k)}}{U_{ij}^{(k)}} + H_i^{(k)} \sum_{j=1}^{n} \frac{1}{U_{ij}^{(k)}} \tag{3.40}$$

$$f'(H_i^{(k)}) = \sum_{j=1}^{n} \frac{1}{U_{ij}^{(k)}} \tag{3.41}$$

$$H_i^{(k+1)} = H_i^{(k)} - \frac{\left[Q_i - \sum_{j=1}^{n} \frac{H_j^{(k)}}{U_{ij}^{(k)}} + H_i^{(k)} \sum_{j=1}^{n} \frac{1}{U_{ij}^{(k)}} \right]}{\sum_{j=1}^{n} \frac{1}{U_{ij}^{(k)}}} \qquad (3.42)$$

Finally, a factor ω that takes values between one and two can be added to improve the convergence:

$$H_i^{(k+1)} = H_i^{(k)}(1-\omega) + \frac{\left[\sum_{j=1}^{n} \frac{H_j^{(k)}}{U_{ij}^{(k)}} - Q_i \right]}{\sum_{j=1}^{n} \frac{1}{U_{ij}^{(k)}}} \omega \qquad (3.43)$$

This method, known as the *Successive Over-Relaxation Method (SOR)*, was rec-ommended by Radojković and Klem (1989). Equation 3.43 shows that the piezo-metric head in node i depends on:

Successive Over-Relaxation Method

- piezometric heads at the surrounding nodes $j = 1$ to n, and
- resistance U of the pipes that connect node i with the surrounding nodes.

The size of the equation matrix in this approach is $n_n \times n_{max}$, where n_n is the total number of nodes and n_{max} is a specified maximum number of pipes/surrounding nodes connected to each node in the system.

The iteration procedure is executed separately for nodes and pipes and consists of internal and external cycles. The preparation steps are:

1) Setting the initial values of the flow in each pipe. This is usually based on the velocity of 1 m/s and a given pipe diameter.
2) Calculating the U values based on the initial flows.
3) Setting the initial values of the piezometric head in each node. As in the case of the Hardy Cross Methods, at least one node must be chosen as a reference (fixed head) node, to allow determination of other nodal heads in the system. The ini-tial head in all the other nodes can be selected in relation to the fixed head value.

The iteration starts in the internal cycle, where the nodal heads are determined by Equation 3.43. The calculation for each node is repeated until $H^{(k+1)} - H^{(k)} < \varepsilon_H$ is satisfied for all the nodes in the system. The nodes with a fixed head represent-ing reservoirs and tanks will be omitted in these iterations.

Thereafter, the flow will be calculated as $Q = \Delta H / U$ for each pipe, in the $(l+1)^{th}$ iteration of the external cycle. If $Q^{(l+1)} - Q^{(l)} > \varepsilon_Q$ for any of the pipes, the internal cycle will have to be repeated using the new U values calculated from the latest flow rates.

The iteration stops when the requested flow accuracy has been achieved for all the pipes or the specified maximum number of iterations has been reached.

Even though this method appears complicated at first glance, it is convenient for computer programming. However, when used for manual calculations it will require lots of time even for a network with very few pipes. A spreadsheet application can reduce this but is by no means an alternative to a full-scale computer programme. Nevertheless, it can serve as a useful tool for better understanding of the principle.

Self-study: Spreadsheet Lesson 4-3 (Appendix 7)

3.5.3 The Gradient Method

Global gradient algorithm

The most widely used to solve the network analysis problem is currently the Gradient Method (also known in literature as the *Global Gradient Algorithm* – GGA), which was first introduced by Todini and Pilati (1987; 1988). The linear system of equations formulated by their approach is solved using the Sparse Matrix Method described by George and Liu (1981). Like the previous approaches, the pipe head-loss calculation (Equation 3.15) is combined with the nodal flow continuity (Equation 3.5). With slightly modified notation in order to capture the entire network, the two equations state, respectively:

$$\Delta E_{ij} = \Delta H_{ij} = H_i - H_j = R_{f,ij} Q_{ij}^n + R_{m,ij} Q_{ij}^2 \tag{3.44}$$

which shows the head loss along pipe ij connecting the nodes i and j, and

$$\sum_{j=1}^{n,i} Q_{ij} - Q_i = 0 \tag{3.45}$$

which shows the balance of flows in node i that is connected with n pipes/surrounding nodes. The total number of pipes ij in the network is n_p while the total number of nodes i is n_n. A few of these nodes (n_s) will have a fixed head (H_0), which denotes water reservoirs and tanks i.e. the sources/supply points. At each iteration of the Gradient Method, the new heads in the remaining nodes will be found by solving the matrix equation:

$$AH = F \tag{3.46}$$

A in the above equation is a Jacobian matrix of the size $(n_n-n_s)\times(n_n-n_s)$, H is a vector of the size n_n-n_s of unknown nodal heads, and F is a vector (n_n-n_s) of what is referred to as right-hand-side terms.

The starting point of the algorithm is the system of equations that in the matrix form looks as follows:

$$\begin{bmatrix} A_{11} & \cdots & A_{12} \\ \cdots & \cdots & \cdots \\ A_{21} & \cdots & 0 \end{bmatrix} \begin{bmatrix} Q \\ \cdots \\ H \end{bmatrix} = \begin{bmatrix} -A_{10}H_0 \\ \cdots \\ q \end{bmatrix} \tag{3.47}$$

The vectors in the above equation represent (in the transposed form):

$Q^T = [Q_1, Q_2, \ldots, Q_{np}]$ (the unknown pipe flows);

$H^T = [H_1, H_2, \ldots, H_{nn\text{-}ns}]$ (the unknown nodal heads);

$H_0^T = [H_1, H_2, \ldots, H_{ns}]$ (the known nodal heads);

$q^T = [q_1, q_2, \ldots, q_{nn\text{-}ns}]$ (the known nodal demands; nodal inflows are assumed to have a negative value).

Furthermore, A_{11} is the diagonal matrix containing the pipe properties as described in Equation 3.44, while A_{12}, A_{21}, and A_{10} are the topological matrices indicating the network connectivity. The following matrix equations apply:

$$\overline{A_{12}} = \begin{bmatrix} A_{12} & \cdots & A_{10} \end{bmatrix} \; ; \; A_{12} = A_{21}^T \tag{3.48}$$

For pipe ij, the matrix term in Equation 3.48 will have the value:

- 1 if the pipe flow enters node i,
- -1 if it leaves node i, and
- 0 if there is no pipe connection to node i.

The above method to express the connectivity can be illustrated in the network from Problem 3-12, after recoding the pipes as:

$1\text{-}2\rightarrow(1)$, $2\text{-}3\rightarrow(2)$, $1\text{-}4\rightarrow(3)$, $2\text{-}5\rightarrow(4)$, $3\text{-}6\rightarrow(5)$, $4\text{-}5\rightarrow(6)$ and $5\text{-}6\rightarrow(7)$.

Taking the final flow directions as the initial ones, the terms in the matrix $\overline{A}_{12}$ look like this:

Node	(demand nodes – A$_{12}$)					(A$_{10}$)	Pipe
	2	3	4	5	6	1 – source	
$\overline{A}_{12} =$	1	0	0	0	0	−1	(1)
	−1	1	0	0	0	0	(2)
	0	0	1	0	0	−1	(3)
	−1	0	0	1	0	0	(4)
	0	1	0	0	−1	0	(5)
	0	0	−1	1	0	0	(6)
	0	0	0	−1	1	0	(7)

Developing all the matrices in Equation 3.47 accordingly, the full system of equations describing the network in Problem 3-12 will look as follows:

$$
\left[
\begin{array}{ccccccc:ccccc}
R_{(1)}|Q_{(1)}|^{n-1} & 0 & 0 & 0 & 0 & 0 & 0 & 1 & 0 & 0 & 0 & 0 \\
0 & R_{(2)}|Q_{(2)}|^{n-1} & 0 & 0 & 0 & 0 & 0 & -1 & 1 & 0 & 0 & 0 \\
0 & 0 & R_{(3)}|Q_{(3)}|^{n-1} & 0 & 0 & 0 & 0 & 0 & 0 & 1 & 0 & 0 \\
0 & 0 & 0 & R_{(4)}|Q_{(4)}|^{n-1} & 0 & 0 & 0 & -1 & 0 & 0 & 1 & 0 \\
0 & 0 & 0 & 0 & R_{(5)}|Q_{(5)}|^{n-1} & 0 & 0 & 0 & 1 & 0 & 0 & -1 \\
0 & 0 & 0 & 0 & 0 & R_{(6)}|Q_{(6)}|^{n-1} & 0 & 0 & 0 & -1 & 1 & 0 \\
0 & 0 & 0 & 0 & 0 & 0 & R_{(7)}|Q_{(7)}|^{n-1} & 0 & 0 & 0 & -1 & 1 \\
\hdashline
1 & -1 & 0 & -1 & 0 & 0 & 0 & 0 & 0 & 0 & 0 & 0 \\
0 & 1 & 0 & 0 & 1 & 0 & 0 & 0 & 0 & 0 & 0 & 0 \\
0 & 0 & 1 & 0 & 0 & -1 & 0 & 0 & 0 & 0 & 0 & 0 \\
0 & 0 & 0 & 1 & 0 & 1 & -1 & 0 & 0 & 0 & 0 & 0 \\
0 & 0 & 0 & 0 & -1 & 0 & 1 & 0 & 0 & 0 & 0 & 0
\end{array}
\right]
\left[
\begin{array}{c}
Q_{(1)} \\ Q_{(2)} \\ Q_{(3)} \\ Q_{(4)} \\ Q_{(5)} \\ Q_{(6)} \\ Q_{(7)} \\ \hline H_2 \\ H_3 \\ H_4 \\ H_5 \\ H_6
\end{array}
\right]
=
\left[
\begin{array}{c}
50 \\ 0 \\ 50 \\ 0 \\ 0 \\ 0 \\ 0 \\ \hline 0.010 \\ 0.022 \\ 0.010 \\ 0.018 \\ 0.015
\end{array}
\right]
$$

The system of equations in 3.47 will be further differentiated as:

$$
\begin{bmatrix}
D & \cdots & A_{12} \\
\cdots & \cdots & \cdots \\
A_{21} & \cdots & 0
\end{bmatrix}
\begin{bmatrix}
dQ \\ \cdots \\ dH
\end{bmatrix}
=
\begin{bmatrix}
dE \\ \cdots \\ dq
\end{bmatrix}
\tag{3.49}
$$

where for the iterations k and k+1:

$D(k,k) = nR|Q_{ij}|^{n-1}$;

$dQ = Q^{(k)} - Q^{(k+1)}$; $dH = H^{(k)} - H^{(k+1)}$;

$dE = A_{11}Q^{(k)} + A_{12}H^{(k)} + A_{10}H_0$; $dq = A_{21}Q^{(k)} + q$.

The solution of Equation 3.49:

$$
\begin{bmatrix}
dQ \\ \cdots \\ dH
\end{bmatrix}
=
\begin{bmatrix}
D & \cdots & A_{12} \\
\cdots & \cdots & \cdots \\
A_{21} & \cdots & 0
\end{bmatrix}^{-1}
\begin{bmatrix}
dE \\ \cdots \\ dq
\end{bmatrix}
\tag{3.50}
$$

after the inversion of the system matrix is further developed as:

$$
\begin{bmatrix}
D & \cdots & A_{12} \\
\cdots & \cdots & \cdots \\
A_{21} & \cdots & 0
\end{bmatrix}^{-1}
=
\begin{bmatrix}
B_{11} & \cdots & B_{12} \\
\cdots & \cdots & \cdots \\
B_{21} & \cdots & B_{22}
\end{bmatrix}
\tag{3.51}
$$

where:

$B_{11} = D^{-1} - D^{-1}A_{12}(A_{21}D^{-1}A_{12})^{-1}A_{21}D^{-1}$;

$B_{12} = D^{-1}A_{12}(A_{21}D^{-1}A_{12})^{-1}$;

$B_{21} = (A_{21}D^{-1}A_{12})^{-1}A_{21}D^{-1}$;

$B_{22} = -(A_{21}D^{-1}A_{12})^{-1}$.

The solution for 3.49 can then be written in simplified form as:

$$dQ = B_{11}dE + B_{12}dq \ ;$$
$$dH = B_{21}dE + B_{22}dq \ .$$

while the final system of equations will look like this:

$$H^{(k+1)} = A^{-1}F \tag{3.52}$$

where $A = A_{21}D^{-1}A_{12}$ and $F = A_{21}Q^{(k)} + q - A_{21}D^{-1}A_{11}Q^{(k)} - A_{21}D^{-1}A_{10}H_0$,

and:

$$Q^{(k+1)} = Q^{(k)} - D^{-1}(A_{11}Q^{(k)} + A_{12}H^{(k+1)} + A_{10}H_0) \tag{3.53}$$

Similar to the Linear Theory, the iterative process starts by calculating the initial flows for assumed velocity. In the next step, the nodal heads are updated by using Equation 3.52, followed by the update of pipe flows by using Equation 3.53. The new flow values will be used to further upgrade the heads, and the interchange of calculated heads and flows in equations 3.52 and 3.53 will continue until the difference between the flows in two consecutive iterations has dropped within the specified error ε_Q for all the pipes.

The Gradient Algorithm is the basis of the hydraulic solver in the widely popular EPANET 2 distribution network modelling software of the U.S. Environmental Protection Agency (used here in the exercises discussed in Appendices 2 – 4, and elaborated on further in Appendix 8). The basic steps of the calculation procedure can also be found in Rossman (2000), described in the following, slightly different way.

The iterative scheme includes the correction of pipe flows using the Newton-Raphson Method that gives the pipe flow in the $(k+1)^{th}$ iteration as follows:

$$Q_{ij}^{(k+1)} = Q_{ij}^{(k)} - \frac{f(Q_{ij}^{(k)})}{f'(Q_{ij}^{(k)})} \tag{3.54}$$

Plugging 3.44 into 3.54 yields:

$$f(Q_{ij}^{(k)}) = R_f Q_{ij}^{n(k)} + R_m Q_{ij}^{2(k)} - \left(H_i - H_j\right)^{(k)} \tag{3.55}$$

$$f'(Q_{ij}^{(k)}) = nR_f Q_{ij}^{n-1(k)} + 2R_m Q_{ij}^{(k)} \tag{3.56}$$

and the flow correction then becomes:

$$\Delta Q_{ij} = \frac{R_f Q_{ij}^n + R_m Q_{ij}^2 - \left(H_i - H_j\right)}{nR_f Q_{ij}^{n-1} + 2R_m Q_{ij}} = y_{ij} - p_{ij}\left(H_i - H_j\right) \tag{3.57}$$

Taking further into consideration different flow directions (+/– signs):

$$p_{ij} = \frac{1}{nR_f \left|Q_{ij}\right|^{n-1} + 2R_m \left|Q_{ij}\right|} \tag{3.58}$$

$$y_{ij} = p_{ij}\left(R_f \left|Q_{ij}\right|^{n} + R_m \left|Q_{ij}\right|^{2}\right) \mathrm{sgn}\left(Q_{ij}\right) \tag{3.59}$$

where $\mathrm{sgn}(Q_{ij}) = 1$ for $Q_{ij} > 0$ and otherwise -1. The diagonal elements of the Jacobian matrix A in Equation 3.46 are composed of the p_{ij} terms, while the off-diagonal (non-zero) elements are $-p_{ij}$. Furthermore, each right-hand-side term will consist of the nodal flow imbalance including a flow connection factor:

$$F_i = \left(\sum_{j=1}^{n,i} Q_{ij} - Q_i\right) + \sum_{j=1}^{n-n_f,i} y_{ij} + \sum_{f=1}^{n_f,i} p_{ij}H_f \tag{3.60}$$

Index f in Equation 3.60 refers to n_f fixed head nodes and the corresponding pipes connecting these to node i.

The essence of the Applied Sparse Matrix Method is in node re-ordering, which allows for filling/computation of the matrix using only non-zero elements, in significantly reduced computational time. Initial pipe flows are assumed without requirements to satisfy the nodal continuity equations; so, using uniform pipe velocity of 1 m/s to calculate them will be sufficient to start the iterations. After the nodal heads have been computed by solving the matrix (Equation 3.46), the new flows are found by correcting the previous flows using Equation 3.57. The iterative process will continue until a particular margin of error has been satisfied. A more detailed description of the matrix composition and handling can further be found both in Todini and Pilati (1988) and Rossman (2000).

In addition to its robustness, the additional advantage of the Gradient Method is that it can handle the change in the status of system components (pumps and valves) without changing the structure of the equation matrix. Rossman (2000) offers further information on how these components are included in the above equations and the iterative process, which is not elaborated on here due to various approaches being possible in the modelling of pump/valve H–Q curves. Nevertheless, a few practical details can be found in Appendix 8.

3.6 Pressure-related demand

It is a known fact that more water is consumed when there is higher pressure in the system, resulting in higher water flows at the outlets. In addition to this, leakage levels are pressure-sensitive, as can be seen from the British experience shown in Figure 3.23.

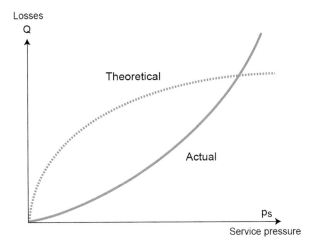

Figure 3.23 Pressure-related leakage

Source: Wessex Water, 1993.

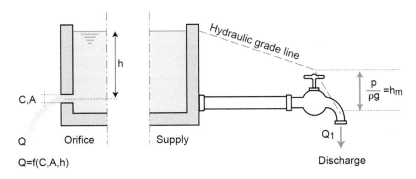

Figure 3.24 Discharge through an orifice

All the calculation procedures explained in the previous sections actually deal with pipe flows against the hydraulic gradients, with the pressures calculated afterwards. As the reference/fixed head (pressure) is set independently from the flows, some error is introduced by neglecting the relation between the demand and pressure in the system. The mathematical relation between these two is quadratic, which can be derived from the analogy with the discharge through an orifice (see Figure 3.24).

The water pressure at the orifice is assumed to be atmospheric and applying the Bernoulli Equation for the cross section just before and just after the orifice leads to the equation:

$$Q = CA\sqrt{2gh} \quad \Leftrightarrow \quad h_m = \xi \frac{Q^2}{A^2 2g} \quad \Rightarrow \quad Q = \frac{1}{\sqrt{\xi}} A\sqrt{2gh_m} \qquad (3.61)$$

where A is the surface area of the orifice and C is a factor (<1) related to its shape. Water depth h above the orifice can be compared with the energy head difference (h ≈ ΔE ≈ ΔH), while the C factor corresponds to the minor loss factor, ξ. Neglecting the friction, Equation 3.61 actually has a format comparable to the term $h_m = R_m Q^2$ shown in Equation 3.15 and further elaborated on by Equation 3.27. Finally, it shows that the residual pressure in water distribution systems is destroyed at the tap, i.e. it has in essence the status of a minor loss ($h_m = p/\rho g$). In reality, applying this logic creates two potential problems:

Demand-driven calculation (DD)

1) Demand-driven (DD) hydraulic calculations will require a correction of the nodal discharges according to the calculated pressure, which may significantly increase the calculation time leading to an unstable iteration procedure. In other words, an input parameter (demand) becomes dependent on an output parameter (pressure).

2) Resistance R_m is, in the case of hundreds of nodes supplying thousands of consumers, virtually impossible to determine in reality. In the best possible case, a general pressure-related diagram may be created from a series of field measurements.

From the Dutch experience, KIWA suggested a linear relation between the pressure and demand for calculations carried out for the assessment of distribution network reliability. The demand is considered independently of the pressure above a particular threshold, which is typically a pressure of 20-30 mwc (Figure 3.25).

Running hydraulic calculations for low-pressure conditions without taking the pressure-related demand into consideration may result in negative pressures in some nodes. This happens if for example the supply head is too low (see Figure 3.26) or the head losses in the system are exceptionally high. Proper interpretation of the results is necessary in this case in order to avoid false conclusions about the system operation.

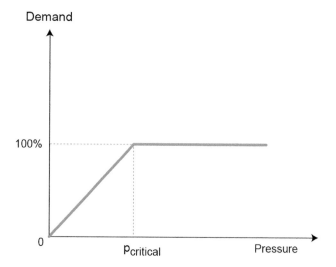

Figure 3.25 Pressure-related demand relation

Source: KIWA, 1993.

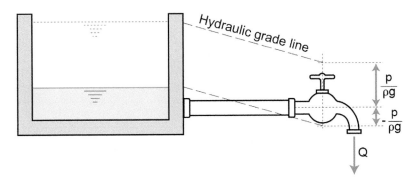

Figure 3.26 Negative pressure as a result of a calculation without pressure-related demand

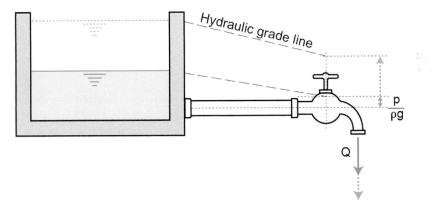

Figure 3.27 Pressure as the result of the calculation with pressure-related demand

Applying the pressure-related demand mode in calculations causes a gradual reduction in the nodal discharges and the hydraulic gradient values, resulting in a slower drop in the reservoir levels, as Figure 3.27 shows.

Tanyimboh *et al.* (2001) describe the pressure-driven demand (PDD) relationship as:

Pressure-driven demand (PDD)

$$H_i = H_i^{min} + K_i Q_i^n \tag{3.62}$$

where H_i represents the actual head at demand node i, H_i^{min} is the minimum head below which the service becomes terminated, K_i is the resistance coefficient for node i, Q_i is the nodal discharge, and n is the exponent that theoretically takes the value of 2.0, as well as usually in practice according to Gupta and Bhave (1996). To determine the unknown value of Q_i for any given nodal head, Equation 3.62 should be rearranged as:

$$Q_i = \left(\frac{H_i - H_i^{min}}{K_i} \right)^{1/n} \tag{3.63}$$

When Q_i equals the required demand, Q_{req}, the value for H_i should equal the desired head, H_{des}, in the node. It is the head that should be available if the demand at that node is to be satisfied in full. Hence:

$$Q_i^{req} = \left(\frac{H_i^{des} - H_i^{min}}{K_i} \right)^{1/n} \Rightarrow \frac{1}{K_i^{1/n}} = \frac{Q_i^{req}}{\left(H_i^{des} - H_i^{min} \right)^{1/n}} \tag{3.64}$$

Finally, substituting K_i in Equation 3.63 yields:

$$Q_i^{avl} = Q_i^{req} \left(\frac{H_i^{avl} - H_i^{min}}{H_i^{des} - H_i^{min}} \right)^{1/n} \tag{3.65}$$

where Q_i^{avl} is the discharge available for the head available at the node (H_i^{avl}). Equation 3.65 considers three possible situations:

1. $H_i^{avl} \leq H_i^{min} \Rightarrow Q_i^{avl} = 0$
2. $H_i^{min} < H_i^{avl} < H_i^{des} \Rightarrow 0 < Q_i^{avl} < Q_i^{req}$
3. $H_i^{avl} \geq H_i^{des} \Rightarrow Q_i^{avl} = Q_i^{req}$

and as such it is used in balancing the flows in the pipes connected to node i. The solution algorithms for solving the system of head equations have been described by Gupta and Bhave (1996). The head-driven simulation will be able to determine the nodes with an insufficient supply. Apart from being a more complex and longer simulation, the key problem here is the correct definition of the values for H_i^{min} and H_i^{des}, i.e. their correlation to the nodal resistance K_i, which describes the nature of the PDD relationship that is essentially empirical. Using this approach, the critical pressure in Figure 3.25 becomes $p_i^{crit}/\rho g = H_i^{des} - H_i^{min}$.

Rossman (2000) describes the concept of so-called *emitter coefficients* by using a similar relationship to Equation 3.62. An emitter is modelled as a setup of a dummy pipe connecting the actual node with a dummy reservoir whose initial head equals the nodal elevation, z. Hence, $H_i^{min} = z_i$ and:

$$Q_i = \frac{1}{K_i^{1/n}} (H_i - z_i)^{1/n} \tag{3.66}$$

Strictly speaking, the K value in Equation 3.66 stands for the resistance of the dummy pipe, but actually it has the same meaning as in equations 3.62 to 3.64. Finally,

$$Q_i = k_i \left(\frac{p_i}{\rho g} \right)^{\alpha}; \quad \frac{p_i}{\rho g} = H_i - z_i; \quad \alpha = 1/n; \quad k_i = \frac{1}{K_i^{\alpha}} \tag{3.67}$$

where k_i stands for the emitter coefficient in node i and α is general emitter exponent with a theoretical value of 0.5.

Emitter coefficients were first introduced to simulate the operation of fire hydrants. In essence, an emitter is a node in which the demand is adjusted based on the actual pressure in the system, following Equation 3.67. The gradual drop in pressure will be followed by a drop in demand but nevertheless, negative pressures may be unavoidable in networks with extremely high differences in elevations (one simple example has been already shown in Figure 3.7). This and a few other deficiencies related to the numerical algorithm using emitters are summarised by Piller and Van Zyl (2009).

Hydraulic calculations taking into consideration the relation between pressures and demands are essential for the analysis of irregular operation scenarios. These include various types of calamities, intermittent supply, water losses through leakage, or calculations of hydrant capacity needed for firefighting, which can also be used for the development of flushing programs. Several approaches and improvements of the PDD calculations including the extension of the Gradient Method can be found in a wide range of literature, for example Ozger and Mays (2003), Ang and Jowitt (2006), Todini (2006), Wu et al. (2006, 2009), and Giustolisi et al. (2008).

Problem 3-13

A pipe $L = 1000$ m, $D = 400$ mm, and $k = 0.5$ mm discharges water from a reservoir with total volume $V_{res} = 2400$ m³, as shown in Figure 3.26; the elevation of the tap is $Z_{tap} = 15$ msl. When the reservoir is full, the water surface elevation is $H_{res} = 20$ msl at a depth of $h_{res} = 6.0$ m, while the pressure on the tap is $p/\rho g = 4.0$ mwc.

Assuming water temperature $T = 10\ °C$:

a) Calculate the loss of reservoir volume after six hours of discharge, and the pressure on the tap at that point in time.
b) Calculate the loss of volume after six hours in intervals of one hour, by taking into consideration the linear dependence between the pressure and discharge on the tap.
c) Repeat the same calculation as under b) by reducing the interval to 30 minutes.

Answers:

a) With pressure on the tap of 4.0 mwc, the corresponding head $H_{tap} = 15 + 4 = 19$ msl and the available head loss along the pipe is $\Delta H = H_{res} - H_{tap} = 20 - 19 = 1$ mwc. Consequently, the pipe hydraulic gradient $S = 1 / 1000 = 0.001$ and, using the spreadsheet hydraulic lesson 1-4, the discharge is calculated as $Q = 74.64$ l/s or 268.69 m³/h. The total loss of volume in six hours is then 1612.14 m³, making the remaining volume $V_{res} = 2400 - 1612.14 = 787.86$ m³. For this volume, the water depth in the reservoir drops from 6.0 m to 787.86 / 2400 x 6.0 = 1.97 m. In the demand-driven calculation, the drop of pressure will be equal to the drop of volume resulting in the negative pressure of $p/\rho g = 4.0 - (6.0 - 1.97) = - 0.03$ mwc.

Lesson 1-4
Maximum Capacity

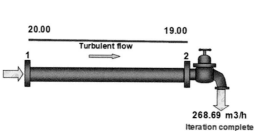

INPUT		OUTPUT	
L (m)	1000	h, (mwc)	1.00
D (mm)	400	u (m²/s)	1.31E-06
k (mm)	0.5	Re (-)	180632
S (-)	0.001	λ (-)	0.0222
T (°C)	10	v (m/s)	0.59
H₂ (msl)	19	Q (l/s)	74.64
Assumption			
v (m/s)	0.59		

20.00   19.00

Turbulent flow

1   2

268.69 m3/h
Iteration complete

b) With the discharge of 268.69 m³/h. The water depth in the reservoir
drops after the first hour from 6.0 m to (2400 – 268.69) / 2400 x 6.0 =
5.33 m. The pressure on the tap also drops but to a lesser extent, due to
intermediate flow reduction. The calculation in each time step follows
a trial-and-error approach in which the selected values of pressure and
discharge of the tap need to match the corresponding hydraulic gradient and the pipe flow. The calculation starts by guessing the pressure
drop on the tap (p/ρg). The head difference H_{res} – H_{tap} is then used
to calculate S and furthermore the pipe flow, Q_{pipe}, using the spreadsheet hydraulic lesson 1-4. The result is final if Q_{pipe} matches the value
of Q_{tap} resulting from the linear relation with the pressure. Complete
results for the time step of one hour are given in the following table.
Q_{err} shows the difference between the calculated pipe flow and the discharge of the tap. The results also show a gradual reduction in the volume and the pressure drop due to the reduction in the drop of discharge,
which reflects real case scenarios. Finally, the reservoir appears to be
approximately half full after 6 hours of discharge.

Time (h)	V_{res} (m³)	H_{res} (msl)	h_{res} (m)	Δh_{res} (m)	H_{tap} (msl)	S (m/km)	Q_{pipe} (m³/h)	p/ρg (mwc)	Q_{tap} (m³/h)	$\Delta p/\rho g$ (mwc)	Q_{err} (m³/h)
0	2400.00	20.00	6.00	–	19.00	0.00100	268.69	4.00	–	–	–
1	2131.31	19.33	5.33	-0.67	18.54	0.00079	238.08	3.54	237.79	-0.46	0.29
2	1893.23	18.73	4.73	-0.60	18.11	0.00062	209.97	3.11	208.91	-0.43	1.06
3	1683.26	18.21	4.21	-0.52	17.73	0.00048	184.02	2.73	183.38	-0.38	0.64
4	1499.24	17.75	3.75	-0.46	17.38	0.00037	160.78	2.38	159.87	-0.35	0.91
5	1338.46	17.35	3.35	-0.40	17.07	0.00028	139.02	2.07	139.05	-0.31	-0.03
6	1199.44	17.00	3.00	-0.35	16.79	0.00021	119.45	1.79	120.24	-0.28	-0.79

c) The same principle was applied by taking the time step of 30 minutes. The results in the table below show that the reservoir is dis-
charging more slowly than in the calculations using a time step of
one hour.

Time (h)	V_{res} (m³)	H_{res} (msl)	h_{res} (m)	h_{res} (m)	H_{tap} (msl)	S (m/km)	Q_{pipe} (m³/h)	$p/\rho g$ (mwc)	Q_{tap} (m³/h)	$\Delta p/\rho g$ (mwc)	Q_{err} (m³/h)
0	2400.00	20.00	6.00	–	19.00	0.00100	268.69	4.00	–	–	–
0.5	2265.66	19.66	5.66	−0.34	18.77	0.00089	253.11	3.77	253.24	−0.23	−0.13
1	2139.10	19.35	5.35	−0.32	18.55	0.00080	239.58	3.55	238.46	−0.22	1.12
1.5	2019.31	19.05	5.05	−0.30	18.34	0.00071	225.30	3.34	224.36	−0.21	0.94
2	1906.66	18.77	4.77	−0.28	18.14	0.00063	211.81	3.14	210.92	−0.20	0.89
2.5	1800.76	18.50	4.50	−0.26	17.95	0.00055	197.47	2.95	198.16	−0.19	−0.69
3	1702.02	18.26	4.26	−0.25	17.77	0.00049	185.93	2.77	186.07	−0.18	−0.14
3.5	1609.06	18.02	4.02	−0.23	17.59	0.00043	173.69	2.59	173.98	−0.18	−0.29
4	1522.21	17.81	3.81	−0.22	17.43	0.00038	162.94	2.43	163.23	−0.16	−0.29
4.5	1440.74	17.60	3.60	−0.20	17.27	0.00033	151.32	2.27	152.48	−0.16	−1.16
5	1365.08	17.41	3.41	−0.19	17.12	0.00029	141.48	2.12	142.41	−0.15	−0.93
5.5	1294.34	17.24	3.24	−0.18	16.98	0.00026	133.77	1.98	133.00	−0.14	0.77
6	1227.46	17.07	3.07	−0.17	16.85	0.00022	122.47	1.85	124.27	−0.13	−1.80

The results for shorter time intervals can be considered to be more accurate because they take the flow reduction into account more continuously. To prove this point, the same calculation has been repeated at more coarse time interval of two hours, which shows that the reservoir is losing water faster than the calculation at one-hour interval shows.

Time (h)	V_{res} (m³)	H_{res} (msl)	h_{res} (m)	h_{res} (m)	H_{tap} (msl)	S (m/km)	Q_{pipe} (m³/h)	$p/\rho g$ (mwc)	Q_{tap} (m³/h)	$\Delta p/\rho g$ (mwc)	Q_{err} (m³/h)
0	2400.00	20.00	6.00	–	19.00	0.00100	268.69	4.00	–	–	–
2	1862.62	18.66	4.66	−1.34	18.06	0.00060	206.56	3.06	205.55	−0.94	1.01
4	1449.50	17.62	3.62	−1.03	17.28	0.00034	153.78	2.28	153.15	−0.78	0.63
6	1141.94	16.85	2.85	−0.77	16.66	0.00019	113.40	1.66	111.51	−0.62	1.89

3.7   Hydraulics of storage and pumps

Reservoirs and pumps are constructed to maintain the energy levels needed for water to reach the discharge points. Factors that directly influence the position, capacity and operation of these components are:

- the topographical conditions,
- location of supply and demand points,
- patterns of demand variation, and
- conveyance capacity of the network.

3.7.1   System characteristics

The conveyance capacity of a pipe with a known length, diameter, roughness factor and slope is described by the *pipe characteristics* diagram. This diagram shows the required heads at the upstream side of the pipe, which enable the supply

Pipe characteristics

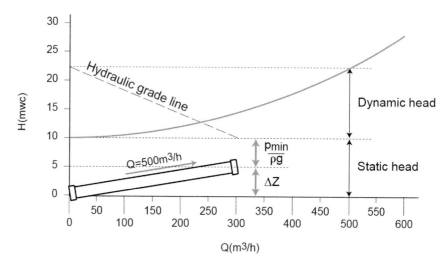

Figure 3.28 Pipe characteristics

of a range of flows while maintaining constant pressure at the pipe end. The total head required for flow Q, in particular, consists of a dynamic and static component (see Figure 3.28).

Dynamic head　The *dynamic head* covers the head losses i.e. the pipe resistance:

$$H_{dyn} = \Delta E = \Delta H = h_f + h_m \tag{3.68}$$

Static head　The *static head* is independent of the flow:

$$H_{st} = \frac{p_{end}}{\rho g} \pm \Delta Z \tag{3.69}$$

where p_{end} stands for the remaining pressure at the pipe end. In design problems where the required head has to be determined for the maximum flow expected in the pipe, the pressure at the end will be fixed at a critical value, i.e. $p_{end} = p_{min}$. *Maintaining the specified minimum pressure at any flow $Q \leq Q_{max}$ will result in the least energy input required for water conveyance at the given pipe characteristics.*

The ΔZ in Figure 3.28 is related to the pipe slope and represents the elevation difference between the discharge point and supply point of the pipe. A positive ΔZ indicates the necessity of pumping while, if there is a negative value, the gravity may partly or entirely be involved in water conveyance. In the example from the figure, the static head $H_{st} = 10$ mwc comprises the minimum downstream pressure head of 5 mwc and 5 m of the elevation difference between the pipe ends. Such a pipe could deliver a maximum flow of 500 m³/h if the head at the supply point was raised up to 22.5 mwc.

The curve for a single (transportation) pipe will be drawn for fixed L, D and k values, and a range of flows covering the demand variations on a maximum consumption day. The minor losses are usually ignored and the curve will be plotted using the results of the friction loss calculation for various flow rates, as explained in Section 3.3.1.

In cases when the diagram in Figure 3.28 represents a network of pipes, it will be called the *system characteristics*. A quadratic relation between the system heads and flows can be assumed in this case: *System characteristics*

$$H_{dyn} = R_s Q^2 \qquad\qquad (3.70)$$

where R_s is the system resistance determined from pressure measurements for various demand scenarios.

The static head of the system will have the same meaning as in Equation 3.69 except that the end in this context means the most critical point i.e. with the lowest expected pressure. This point can exist at the physical end of the system, far away from the source, but can also be within the system if located at a high elevation (thus, resulting in a high ΔZ).

The system/pipe curves change their shape as the head loss varies resulting from modification of the pipe roughness, diameter or length. This can be a con-sequence of:

- pipe ageing, and/or
- system rehabilitation/extension.

The curve becomes steeper in all cases as the head loss increases, which results in the reduction of the conveying capacity. The original capacity can be restored

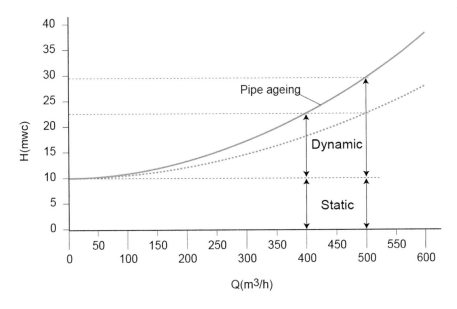

Figure 3.29 Capacity reduction in the system.

by laying new pipes of a larger diameter, or in a parallel arrangement, as discussed in Section 3.3.5. Alternatively, more energy, i.e. a higher head at the supply side, will be needed in order to meet the demand. The example in Figure 3.29 shows how the pipe from Figure 3.28 requires an upstream head of nearly 30 mwc after it loses its initial conveyance due to ageing, in order to keep the same supply of 500 m³/h. Maintaining the initial supply head of 22.5 mwc would otherwise cause a reduction in supply to 400 m³/h.

Self-study: Spreadsheet Lessons 1-6 and 2-1b (Appendix 7)

3.7.2   Gravity systems

In the case of *gravity* systems, the entire energy needed for water flow is provided from the elevation difference ΔZ. The pressure variation in the system is influenced exclusively by the demand variation (see Figure 3.30). Hence:

$$\Delta Z = H_{dyn} + H_{st} = \Delta H + \frac{P_{end}}{\rho g} \tag{3.71}$$

For known pressure at the end of the system, the maximum capacity can be determined from the system curve, as shown in Figure 3.31. The figure shows that the lower demand overnight causes a smaller head loss and therefore the minimum pressure in the system will be higher than during the daytime when the demand and head losses are higher. In theory, this has implications for the value of the static head that changes with the variable minimum pressure. *The static head used for design purposes is always fixed based on the minimum pressure that has to be maintained in the system during the maximum consumption hour.*

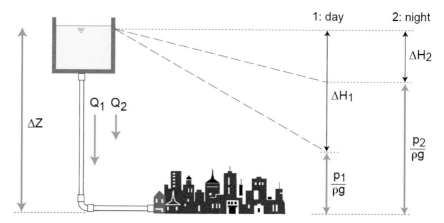

Figure 3.30 Gravity system: regular supply.

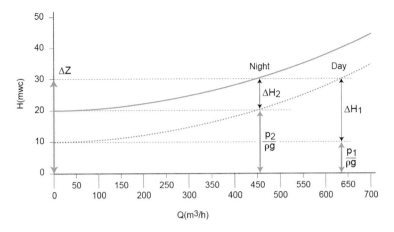

Figure 3.31 System characteristics: regular operation

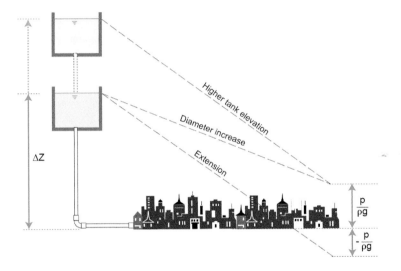

Figure 3.32 Gravity system: network extension

When an area that is to be supplied from a single source starts to grow considerably, demand increases and longer pipe routes can lead to a pressure drop in the network that affects the newly-constructed areas. In theory, this problem can be solved by enlarging the pipes and/or elevating the reservoir (figures 3.32 and 3.33). Nonetheless, the latter is often impossible due to the fixed position of the source and additional head will probably have to be provided by pumping.

If the system is supplied from more than one side, the storage that is at the higher elevation will normally provide more water i.e. the coverage of the larger part of the distribution area. The intersection between the hydraulic grade lines shows the line of separation between the areas covered by different reservoirs, which is referred to as the so-called *zero line* (Figure 3.34).

Zero line

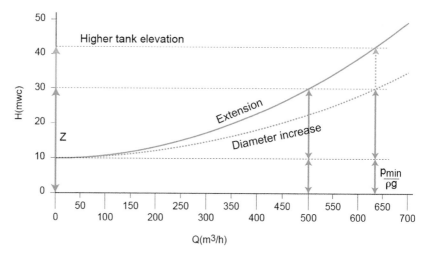

Figure 3.33 System characteristics: network extension

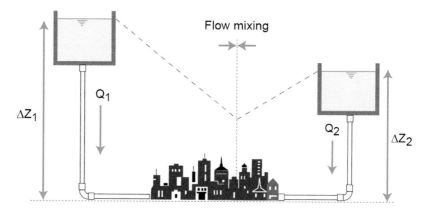

Figure 3.34 Gravity system: supply from two sides.

Hydraulic conditions in the vicinity of the zero line are unfavourable because:

- the pressure is lower than in other parts of the network, and
- the flow velocities are also low, leading to water stagnation and potential water quality problems.

Problem 3-14

For the gravity system shown in the figure below, find the diameter of the pipe L = 2000 m that can deliver a flow of 6000 m³/h with a pressure of 40 mwc at the entrance to the city. Absolute roughness of the pipe can be assumed at k = 1 mm and the water temperature equals 10° C.

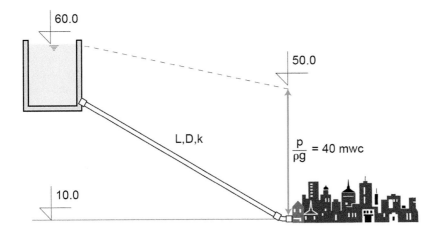

What will the increase in capacity of the system be if the pressure at the entrance to the city drops to 30 mwc?

Answers:

At the elevation of Z = 10 msl and the pressure of p/ρg = 40 mwc, the piezo-metric head at the entrance to the city becomes H = 50 msl. The surface level/piezometric head of the reservoir is 10 meters higher, and this difference can be utilised as friction loss. The hydraulic gradient then becomes S = h_f / L = 10 / 2000 = 0.005. From the hydraulic tables in Appendix 6, for k = 1 mm and T = 10° C:

Discharge flows (l/s), k = 1 mm, S = 0.005

D (mm)	Q (m³/h)
900	4772.0
1000	6292.6

If the pressure drops to p/ρg = 30 mwc, the available friction loss increases to h_f = 20 mwc and therefore S = 0.01. From the same tables, for D = 1000 mm, the flow that can be supplied for the given hydraulic gradient increases to Q = 8911.1 m³/h.

The same results can be obtained by the calculation procedures demonstrated in problems 3-8 and 3-7 in sections 3.3.3 and 3.3.2, respectively.

Self-study: Workshop problem 1.5.1 (Appendix 1)
Spreadsheet Lessons 5-1 to 5-5 (Appendix 7)

3.7.3 Pumped systems

Pump lift

Suction and discharge pipe

In pumped systems, the energy needed for water conveyance is obtained from the pump operation. This energy, generated by the pump impeller, is usually expressed as a head of water column (in mwc) and is called the *pumping head* (or *pump lift*), h_p. This represents the difference between the energy levels at the pump entrance i.e. at the *suction pipe* and at the pump exit, i.e. at the *discharge* (or *pressure*) *pipe* (Figure 3.35).

In the case of a single pump unit, the higher the pumping head h_p is, the smaller the pumped flow Q will be, as shown in Figure 3.36. For a combination of Q-h_p values, the power N (kW) required to lift the water is calculated as:

$$N = \rho g Q h_p \tag{3.72}$$

Pump discharge

where Q (m³/s) is the *pump discharge*. The power required to drive the pump will be higher, due to energy losses in the pump:

$$N_p = \frac{\rho g Q h_p}{\eta_p} \tag{3.73}$$

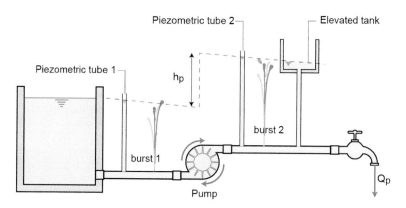

Figure 3.35 Pump operation

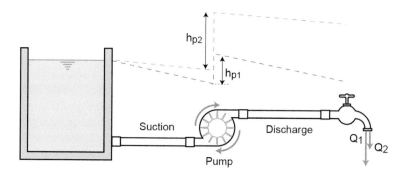

Figure 3.36 Pumping head and flow

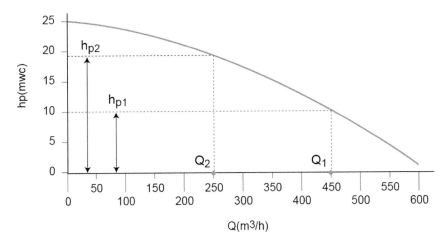

Figure 3.37 Pumping characteristics

where η_p is the pump efficiency dependant on the pump model and working regime. Finally, the power required for the pump motor will be:

$$N_m = \frac{N_p}{\eta_m}$$ (3.74)

where η_m indicates the motor efficiency.

The hydraulic performance of pumps is described by the *pump characteristics*. This diagram shows the relation between the pump discharge and delivered head (Figure 3.37).

Pump characteristics

For centrifugal pumps, a very good approximation of the pump curve is achieved by the following equation:

$$h_p = aQ^2 + bQ + c$$ (3.75)

where factors a, b and c depend on the pump model and flow units. The alternative equation can also be used:

$$h_p = c - aQ^b$$ (3.76)

This equation allows for the pump curve definition with a single set of Q-h_p points (Rossman, 2000). These are known as the *duty flow (Q$_d$)* and *duty head (H$_d$)* and indicate the optimal operational regime of the pump, i.e. the one in which the maximum efficiency η_p will be achieved. As a convention, for exponent b = 2:

Duty flow and head

$$h_{p(Q=0)} = c = \frac{4}{3}H_d; \quad Q_{(h_p=0)} = 2Q_d \quad \Rightarrow \quad a = \frac{1}{3}\frac{H_d}{Q_d^2}$$ (3.77)

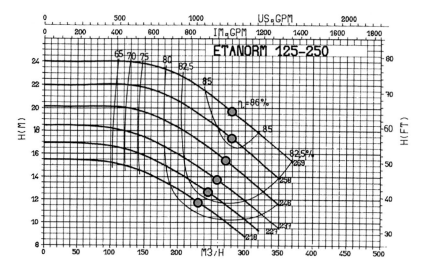

Figure 3.38. Pump characteristics from a manufacturer's catalogue

Adapted from: KSB, 1990.

Pump manufacturers regularly supply pump characteristics diagrams for each model; a typical format showing a range of impeller diameters and efficiencies η_p, is given in Figure 3.38. The dots in the figure indicate the points of duty flow and duty head.

Following the discussions in Section 3.7.1, the pumping head required at the supply side of the system to maintain a particular minimum pressure at its end will be:

$$h_p^{req} = H_{dyn} + H_{st} = \Delta H + \frac{P_{min}}{\rho g} \pm \Delta Z \tag{3.78}$$

This required head is normally higher during the daytime than overnight, resulting from higher demand i.e. higher head losses. Operating the same pump (curve) over 24 hours will therefore result in the opposite effect: low heads during the daytime and high heads overnight (Figure 3.39). In addition, using a single pump in a pumping station is unjustified for reasons of low reliability, high-energy consumption/low efficiency and problematic maintenance. In practice, several pumps are commonly combined in one pumping station.

Flows and pressures that can be delivered by pump operation are determined from the system and pump characteristics. The intersection of these two curves, *Working point* the *working point*, indicates the required pumping head that provides the flow and the static head, as shown on the graph in Figure 3.40.

As in the case of the gravity supply discussed in the previous section (3.7.2), the pressure variation in the distribution area has implications for the value of the static head in this case as well. For design purposes, however, *the working point obtained from the system characteristics plotted at the lowest static head (i.e. the lowest pressure required in the system) will be used to determine the maximum pump capacity needed to satisfy the required service level.*

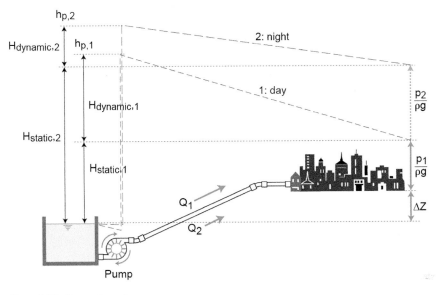

Figure 3.39 Operation of one pump

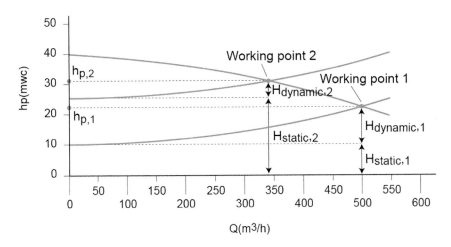

Figure 3.40 Operation of one pump: daytime and night time flows

The maximum pumping capacity may vary over time. Ageing of the pump impeller, pipe corrosion, increase in leakage, etc. will cause a reduction in the maximum flow that can be delivered by the pump while maintaining the same static head (Figure 3.41).

Decisions on the number and size of pumps in a pumping station are made with the general intention of keeping the pressure variations in the system at the lowest possible level in order to minimise the required pumping energy. For this reason, several pumps connected to the same delivery main can be installed in

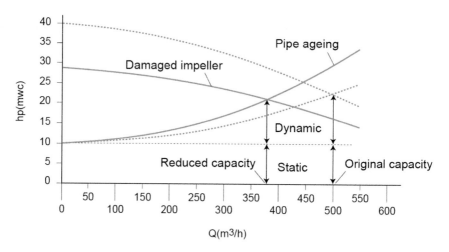

Figure 3.41 Operation of one pump: flow reduction

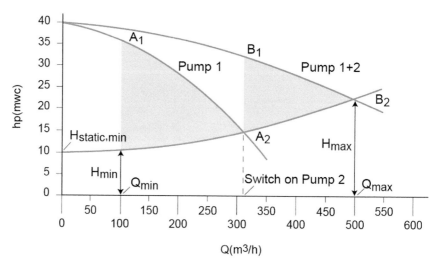

Figure 3.42 Equal pumps in a parallel arrangement

parallel. Their operation will be represented by a composite pump curve, which is obtained by adding the single pump discharges at the same pumping head. Hence, for n pumps:

$$h_p = h_1 = h_2 = h_3 = ... = h_n \; ; \; Q_p = Q_1 + Q_2 + Q_3 + ... + Q_n$$

Figure 3.42 shows the operation of two equal pumping units in a parallel arrangement. The system should preferably operate at any point along the curve A_1-A_2-B_1-B_2, between the minimum and maximum flows. The shaded area in

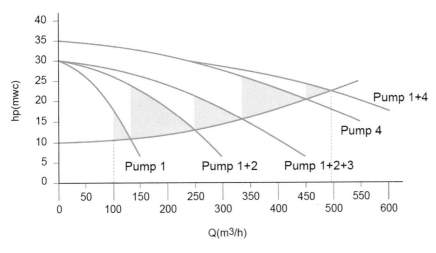

Figure 3.43 Various pump sizes in a parallel arrangement

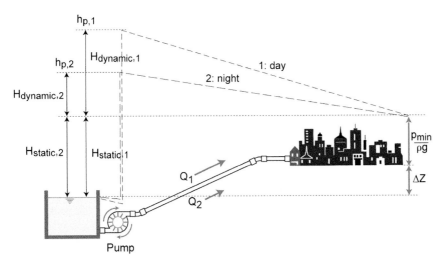

Figure 3.44 Operation of variable speed pumps

the figure indicates excessive pumping, which is unavoidable when fixed-speed pumps are used. A properly selected combination of pump units should minimise this area. This is often achieved by installing pumps with different capacities; the example in Figure 3.43 shows the combination of three equal units (1-3) with one stronger pump (4).

Introducing variable speed pumps can completely eliminate the excessive head. The flow variation is met in this case by adjusting the impeller rotation, keeping the discharge pressure constant (Figure 3.44).

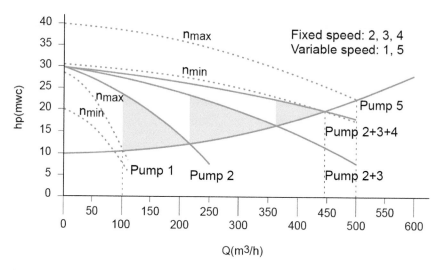

Figure 3.45 Combined operation of variable and fixed speed pumps

The pump characteristic diagram will consist of a family of curves for various pump frequencies, n (rpm). The relation between the various pumping heads and flows of any two curves is proportional to the frequencies as follows:

$$\frac{Q_2}{Q_1} = \frac{n_2}{n_1}; \frac{h_{p,2}}{h_{p,1}} = \left(\frac{n_2}{n_1}\right)^2 \tag{3.79}$$

Nevertheless, one variable speed unit alone can scarcely cover the entire range of flows and therefore several units in parallel will be used. Variable speed pumps can also be combined with fixed speed pumps, controlling only the peak flows (Figure 3.45).

When there are large pressure variations in the system, the pumps have to be installed in a serial arrangement. In this case the total head is equal to the sum of heads for each pump. Figure 3.46 shows the curves for two pumps in operation. For n equal units:

$$h_p = h_1 + h_2 + h_3 + \ldots\ldots + h_n \; ; \; Q_p = Q_1 = Q_2 = Q_3 = \ldots\ldots = Q_n$$

In addition to the discussion at the end of Section 3.7.2, pumping from more than one supply point will cause similar problems as with gravity systems in areas where the water from different sources is mixed. However, modifying the pump regimes can shift the zones of minimum pressure, as shown in Figure 3.47.

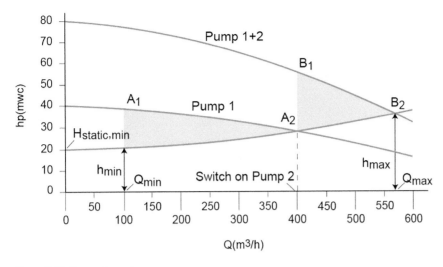

Figure 3.46 Pumps in series

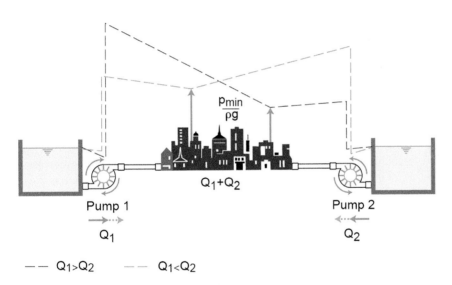

Figure 3.47 Pumping from two sources

Problem 3-15

For the pumped system shown in the figure below, determine the required pumping head to deliver flow $Q = 4000$ m³/h through pipe $L = 1200$ m and $D = 800$ mm, while maintaining the pressure of 50 mwc at the entrance to the city. Absolute roughness of the pipe can be assumed at $k = 0.5$ mm and the water temperature equals 10° C.

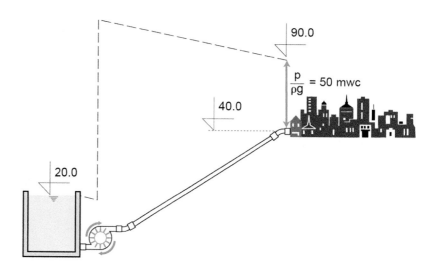

Find the equation of the pump curve using formulas 3.76 and 3.77 assuming the above operation happens at maximum pump efficiency. What will be the pressure at the entrance to the city when there is a demand growth of 25%?

Answers:

At the entrance to the city, elevation $Z = 40$ msl. With a required pressure of $p/\rho g = 50$ mwc, the piezometric head becomes $H = 90$ msl. As the surface level/piezometric head of the reservoir is set at 20 meters, the total static head $H_{st} = 90 - 20 = 70$ mwc. The losses between the reservoir and the pump can be ignored.

For the following parameters: $Q = 4000$ m³/h, $L = 1200$ m, $D = 800$ mm, $k = 0.5$ mm and $T = 10°$ C, the pipe friction loss will be calculated as follows:

$$v = \frac{4Q}{D^2\pi} = \frac{4\times4000}{0.8^2\times3.14\times3600} = 2.21\,m/s$$

For temperature $T = 10$ °C, the kinematic viscosity from Equation 3.22, $\upsilon = 1.31\times10^{-6}$ m²/s. The Reynolds number takes the value of:

$$Re = \frac{vD}{\upsilon} = \frac{2.21\times0.8}{1.31\times10^{-6}} = 1.4\times10^6$$

and the friction factor λ from Barr's Equation equals:

$$\lambda = \frac{0.25}{\log^2\left[\dfrac{5.1286}{Re^{0.89}} + \dfrac{k}{3.7D}\right]} = \frac{0.25}{\log^2\left[\dfrac{5.1286}{(1.4\times10^6)^{0.89}} + \dfrac{0.5}{3.7\times800}\right]} \approx 0.018$$

The friction loss from the Darcy-Weisbach Equation can be determined as:

$$h_f = \frac{\lambda L}{12.1 D^5} Q^2 = \frac{0.018 \times 1200}{12.1 \times 0.8^5} 0.35^2 \approx 7\, mwc$$

The total required pumping head is therefore $h_p = H_{st} + H_{dyn} = 70 + 7 = 77$ mwc. Given the maximum pumping efficiency, this is also the duty head at the duty flow of 4000 m³/h and in Equation 3.77:

$$a = \frac{1}{3} \frac{H_d}{Q_d^2} = \frac{77}{3 \times 4000^2} = 1.604 \times 10^{-6} \text{ and } c = \frac{4}{3} H_d = \frac{4}{3} \times 77 = 102.67$$

Hence, the pumping curve can be approximated with the following equation (exponent b = 2):

$$h_p = c - aQ^b \approx 103 - 1.6 \times 10^{-6} Q^2$$

If demand grows by 25% i.e. to 5000 m³/h, the pumping head that can be provided will be:

$$h_p = 103 - 1.6 \times 10^{-6} 5000^2 \approx 63\, mwc$$

The friction loss calculated using the same method as above is going to increase to approximately 11 mwc, leading to a residual pressure at the entrance to the city of p/ρg = 20 + 63 − 11 − 40 ≈ 32 mwc.

Self-study: Workshop problem 1.5.2 (Appendix 1)
                Spreadsheet Lessons 6-1 to 6-5 (Appendix 7)

3.7.4 Combined systems

Consumers in combined systems are partly supplied by gravity and partly by pumping.
    Three basic concepts can be distinguished:

1) The water is pumped from a reservoir into the distribution area (tank-pump-network).
2) The water is pumped to a reservoir and thereafter supplied by gravity (pump-tank-network).
3) Pump and reservoir are at the opposite sides of the distribution area (pump-network-tank).

Tank-Pump-Network

This scheme is suitable for relatively flat terrains where a favourable location for the reservoir is difficult to find, either due to insufficiently high elevations or because of a large distance from the distribution area.

Essentially this is the same concept as that of direct pumping, except that the required pumping head can be reduced on account of the elevation difference in the system. Hence:

$$h_p + \Delta Z = H_{dyn} + H_{st} = \Delta H + \frac{P_{end}}{\rho g} \tag{3.80}$$

Both the dynamic and static head are supplied partly by gravity and partly by pumping, depending on the elevation difference and the pressure at the end of the system (figures 3.48 and 3.49).

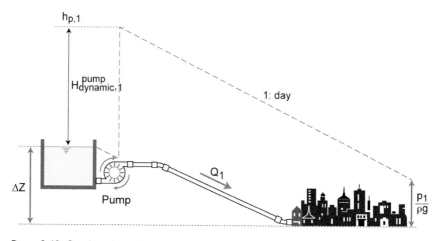

Figure 3.48 Combined supply by gravity and pumping: daytime flows

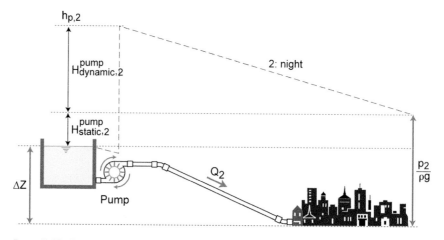

Figure 3.49 Combined supply by gravity and pumping: night time flows

Pumping stations need not necessarily be located at the supply point. When positioned within the system, they are commonly called *booster stations*. Such a layout is attractive if high pressures are to be avoided (Figure 3.50).

Booster stations

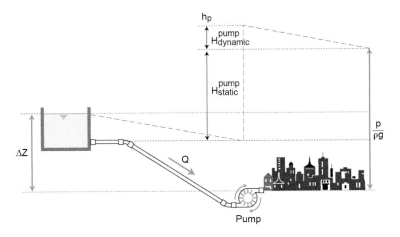

Figure 3.50 A booster station

Pump-Tank-Network

This scheme is typical for hilly terrains. When pumps deliver water to the reservoir, the static head will only comprise the elevation head ΔZ, which equals the elevation difference between the surface levels in the two reservoirs (Figure 3.51). Thus:

$$h_p = H_{dyn} + H_{st} = \Delta H + \Delta Z \tag{3.81}$$

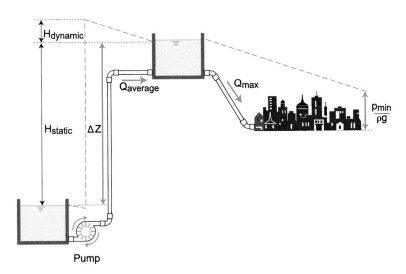

Figure 3.51 Gravity supply supported by pumping

The advantages of this scheme are:

- stable operation of the pumping station, and
- a buffer supply capacity in the case of pump failure.

A similar hydraulic pattern is valid if water towers are put into the system (Figure 3.52). However, their predominant role is to maintain stable operation of the pumps, rather than to provide buffer volumes or large balancing volumes.

While supplying tanks, the pumps often operate automatically, based on monitoring of water levels in the reservoirs. Pump throttling may be required in order to adjust the flow. The effects on the system characteristics are shown in Figure 3.53.

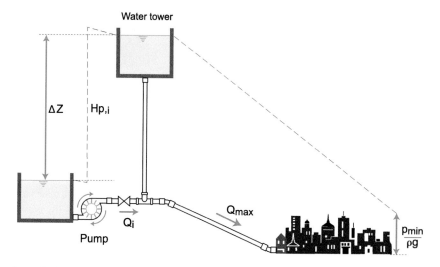

Figure 3.52 Combined supply by pump and water tower

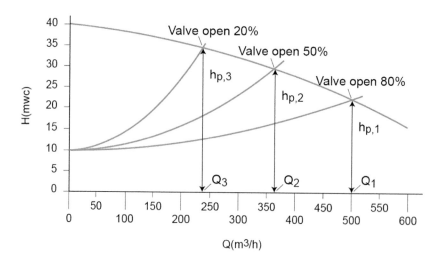

Figure 3.53 Effects of pump throttling on system characteristics

Pump-Network-Tank

This scheme is predominantly applied to distribution networks located in valleys. During the maximum supply conditions, both the pump and reservoir will supply part of the distribution area (Figure 3.54).

If the only source of supply is close to the pumping station, that source will also be used to refill the volume of the tank. This is usually done overnight when the demand in the area is low (Figure 3.55).

A tank operating in this way functions as a kind of *counter tank* to the one at the source. Depending on its size and elevation, it can balance the demand variation   *Counter tank*

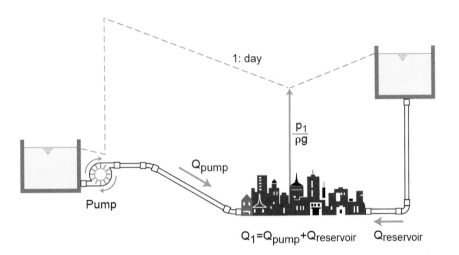

Figure 3.54 Counter tank: daytime flows

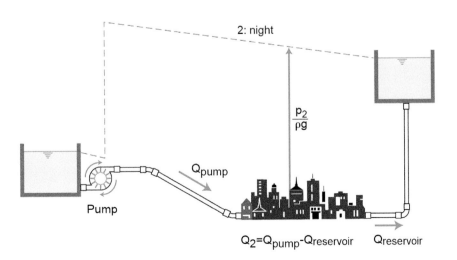

Figure 3.55 Counter tank: night time flows

in the system, either partly or completely. In the latter, the pumping station operates at constant (average) capacity ($Q_{pump} = Q_{average}$).

Self-study: Workshop problems 1.5.3 to 1.5.7 (Appendix 1)
                       Spreadsheet Lessons 7-1 to 7-4 (Appendix 7)

Chapter 4

The design of water transport and distribution systems

The design of water transport and distribution systems consists of two parts: hydraulic and engineering. The main parameters considered in hydraulic design have been discussed in previous chapters. Apart from sufficient flows, pressures and velocities, a well-designed system should fulfil the following additional requirements:

- minimised operational costs in regular supply conditions,
- reasonable supply during irregular situations (power/pump failure, pipe burst, fire events, system maintenance, rehabilitation or reconstruction), and
- flexibility with respect to future extensions.

Keeping the hydraulic parameters within an acceptable range cannot by itself fulfil these requirements. Equally important are what are referred to as *engineering (non-hydraulic)* design criteria, such as: *Engineering design criteria*

- the selection of durable pipe materials, joints, fittings and other appurtenances,
- setting up a network of valves whereby parts of the network can quickly be isolated, and
- providing easy access to the vital parts of the system.

Respecting both the hydraulic and engineering design criteria guarantees satisfactory operation of the system throughout the entire design period.

4.1 The planning phase

Commissioning a water distribution system means a huge investment with far-reaching implications for the development of the area that will be covered by the network. To avoid major mistakes, starting with a good plan is a meaningful preparatory step before the detailed design considerations take place. The planning phase has to answer the following questions (Pieterse, 1991):

1) Is the project feasible?
2) What is the best global approach?
3) What are the estimated costs?
4) What is the required timescale for execution?

Looking for appropriate answers to these questions is often a complex assignment involving experts with different profiles. Hence, organizing the work effectively is an essential element of the planning. The job normally starts by establishing a project management team with the following main tasks:

- a project review,
- a survey of required expertise and equipment,
- the securing of cooperation between involved organisations, and
- the setting of project objectives with respect to time, costs and quality.

Before thinking about any possible solution, any existing information and ideas about the long-term physical planning objectives of the distribution system should be explored. The main strategy of the long-term development of the region is usually stipulated in documents prepared at governmental level. Based on these plans, more specific analyses related to the aspects of water supply will lead to a number of concept solutions. These alternatives are then discussed and evaluated by the studies that form the actual essence of the design (identification report, feasibility study, master plan). Apart from global recommendations on how to approach the design, the outcome of these studies will result in the more detailed organization of the project, such as:

- division of the project into smaller parts,
- definition of project phases (in terms of time), and
- estimates of costs and time necessary for the execution.

Approving these steps and arranging finance are preconditions for starting the design phase.

The conclusions are always made with a safety margin in the planning phase. This is logical, as a period of 20 to 30 years is long enough to include unforeseen events arising from political problems, natural disasters, epidemics, and other (not always negative) factors distorting normal population growth. It is therefore wise to develop water distribution facilities in stages, following the actual development of the area. This principle allows the gradual accumulation of the finance and an intermediate evaluation as well as adaptation of the design if the actual development deviates from the original planning. Thus, the planning phase is never fully completed before the design and execution phases begin.

4.1.1 The design period

Design period

Various components of the distribution system are designed for a particular period of time called the *design period*. During this period, the capacity of the component should be adequate unless the actual water demand differs from the forecast, as Figure 4.1 shows.

Technical lifetime

The *technical lifetime* of a system component represents the period during which it operates satisfactorily in a technical sense. The suggested periods for the

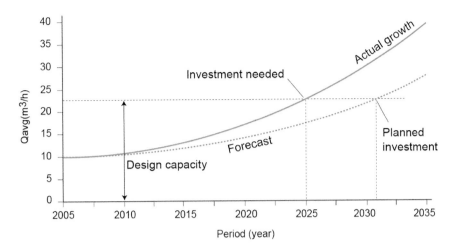

Figure 4.1 Demand forecast

Table 4.1 Technical lifetime of distribution system components

Component	Period (years)
Transmission mains	30 – 60
Distribution mains	30 – 80
Reservoirs	20 – 80
Pumping station - facilities	20 – 80
Pumping station - equipment	15 – 40

main distribution system components shown in Table 4.1 indicate a wide range which mostly depends on the appropriateness of the choice and the way in which the component has been maintained.

The *economic lifetime* represents the period of time for which the component   *Economic lifetime* can operate before it becomes more costly than its replacement. This lifetime is never longer than the technical lifetime; very often it is much shorter. Its estimation is complex and depends on aspects such as operation and maintenance costs, technological advancement, and interest rates.

In practice, the design period is often the same as the economic lifetime. Moreover, a uniform design period will be chosen for all the components; design periods of 20-25 years are typical for distribution systems. An exception is mechanical equipment in pumping stations, which has a lifetime of 10-15 years. Although water companies are sometimes able to successfully maintain pumps operating for longer than 30 years, or pipes with low corrosion that are older than 70 years, experience shows that design periods rarely exceed 30 years. Design periods shorter than 10 years are uneconomical and therefore undesirable.

4.1.2 Economical aspects

The economical comparison of design alternatives is a key element of the final choice; at the same time this is the most subjective part of the whole project.

For practical reasons, the alternatives will be compared within the same design period for all the components, although the most economical design period may differ for individual components. The important factors that influence the most economical design period are:

- interest rates,
- inflation rates,
- energy prices,
- water demand growth, and
- the 'scale' economy.

The 'scale' economy is an approach where investment costs are established in relation to the main properties of the system component. This is possible if the water supply company, or a number of neighbouring companies, have kept sufficient records of relevant costs.

First cost

For instance, the *first cost* (FC) of concrete reservoirs can be calculated as: a × V^n, where V is the tank volume in m³, and a and n are the factors depending on local conditions. A similar relation can be used for pumping stations taking the maximum capacity Q instead of volume V into consideration. Furthermore, linear or exponential relations can be adopted for transmission lines as, for instance, FC = a × D, or FC = b + c × D^n, with D representing the pipe diameter, say in millimetres.

Present/annual worth

A preliminary cost comparison of the considered design alternatives can be carried out using the *present worth* (present value) or the *annual worth* method.

By the present worth method, all actual and future investments are calculated back to a reference year, which in general is the year of the first investment. The alternative with the lowest present value offers the most economical solution. The basic parameter in the calculation is the *single present worth factor*, $p_{n/r}$:

Single present worth factor

$$p_{n/r} = \frac{1}{s_{n,r}} = \frac{1}{(1+r)^n} \tag{4.1}$$

Single compound amount factor

where $s_{n/r}$ is the *single compound amount factor*, which represents the growth of the present worth PW after *n* years with a *compounded interest rate* of r. The present worth of the future sum F then becomes PW = F × $p_{n/r}$.

According to the annual worth method, a present principal sum P is equivalent to a series of n end-of-period sums A, where:

$$A = P \frac{r(1+r)^n}{(1+r)^n - 1} = P \times a_{n/r} \tag{4.2}$$

Annuity

In the above equation, $a_{n/r}$ represents the *capital recovery factor (annuity)*. When the present worth is calculated as PW = A / $a_{n/r}$, the 1 / $a_{n/r}$ is called the *uniform present worth factor*.

Use of an *ideal interest rate* i in equations 4.1 and 4.2, instead of the true interest rate r, allows the impact of inflation to be taken into account. Factor f in Equation 4.3 represents the *annual inflation rate*.

$$i = \frac{r - f}{1 + f} \tag{4.3}$$

The Theory of Engineering Economy offers more sophisticated cost evaluations that can be further studied in appropriate literature; for further information refer for instance to De Garmo *et al.* (1993).

The most economical alternative usually becomes obvious after comparisons between the investment and operational costs and their effects on the hydraulic performance of the component/system. The following is a demonstration of the simplified principle to evaluate the investment and operation and maintenance (O&M) costs for a trunk main.

A pipe conveys flow Q (in m³/s) while generating head loss ΔH (mwc). The cost of energy EC (kWh) wasted over time T (hours) can be calculated as:

$$EC = \frac{\rho g Q \Delta H}{1000 \times \eta} T \times e \tag{4.4}$$

where e is the unit price (per kWh) of the energy needed to compensate for the pipe head loss. By supplying this energy using a pump, the annual costs of the energy wasted per metre length of the pipe become:

$$EC = \frac{9.81 \times 24 \times 365 \times Q\Delta H}{3600 \times L} \frac{e}{\eta} \approx 24 \times Q \frac{e}{\eta} \frac{\Delta H}{L} \tag{4.5}$$

where Q is the average pump flow in m³/h, and η is the corresponding pumping efficiency. Substituting the hydraulic gradient, ΔH/L by using for example the Darcy-Weisbach Equation, the energy cost per annum will be (assuming the friction factor λ is equal to 0.02):

$$EC = 24 \times Q \frac{e}{\eta} \frac{0.02 \times Q^2}{12.1 \times D^5 \times 3600^2} \approx 3 \times 10^{-9} \frac{e}{\eta} \frac{Q^3}{D^5} \tag{4.6}$$

where D is the pipe diameter expressed in metres (and Q in m³/h). By adopting a linear proportion between the pipe diameter and its cost, the total annual costs including investment and operation of the pipe are:

$$A = a \times D \times a_{n/r} + 3 \times 10^{-9} \frac{e}{\eta} \frac{Q^3}{D^5} \tag{4.7}$$

Equation 4.7 has the optimum solution if δA/δD = 0:

$$a \times a_{n/r} = 5 \times 3 \times 10^{-9} \frac{e}{\eta} \frac{Q^3}{D^6} \tag{4.8}$$

which finally results in the most economical diameter:

$$D = 0.05\sqrt{Q} \sqrt[6]{\frac{e}{\eta a_{n/r} a}}$$

(4.9)

Equation 4.9 considers fixed energy costs and water demand over the design period. The growth of these parameters should also normally be taken into account.

Figure 4.2 shows the same approach visually. The diagram in this figure compares investment and operational costs calculated for a range of possible diameters. The larger diameters will obviously be more expensive while generating lower friction losses i.e. generating the lower energy costs. The minimum of the curve summarising these two costs pinpoints the most economical diameter, in this case of 300 mm.

Problem 4-1

A loan of US$ 5,000,000 has been obtained for reconstruction of a water distribution system. The loan has an interest rate of 6 % and repayment period of 30 years. According to alternative A, 40 % of this loan will be invested in the first year and 30 % in years two and three, respectively. Alternative B proposes 60 % of the loan to be invested in year one and the rest in year 10. Which of the two alternatives is cheaper in terms of investment? Calculate the annual instalments if the repayment of the loan starts immediately. What will the situation be if the repayment of the loan starts after ten years?

Answers:

The present worth for both alternatives will be calculated for the beginning of the period. In alternative A:

$$PW_A = \sum_{i=1}^{3} F_i P_{in/6} = 5,000,000 \times \left[\frac{0.4}{(1+0.06)^1} + \frac{0.3}{(1+0.06)^2} + \frac{0.3}{(1+0.06)^3} \right]$$

$$= 4,481,216\ US\$$$

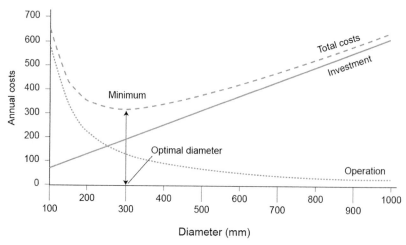

Figure 4.2 Cost comparison of the optimum diameter

while for alternative B:

$$PW_B = \sum_{i=1}^{2} F_i p_{in/6} = 5,000,000 \times \left[\frac{0.6}{(1+0.06)^1} + \frac{0.4}{(1+0.06)^{10}} \right]$$

$$= 3,946,978 \ US\$$$

Due to the postponed investments, alternative B appears to be more cost effective. The annuity calculated from Equation 4.2 for a repayment period of 30 years and interest rate of 6 % becomes:

$$a_{30/6} = \frac{0.06 \times (1+0.06)^{30}}{(1+0.06)^{30} - 1} = 0.0726$$

leading to 30 annual instalments of 0.0726 × 4,481,216 = 325,336 US\$ in the case of alternative A, and 0.0726 × 3,946,978 = 286,550 US\$ for alternative B.

If the repayment of the loan is delayed for 10 years i.e. stretches over 20 years, the calculated annuity becomes:

$$a_{20/6} = \frac{0.06 \times (1+0.06)^{20}}{(1+0.06)^{20} - 1} = 0.0872$$

For the same schedule of investments, the present value in year 10 in alternative A becomes:

$$PW_{A,10} = 5,000,000 \times \left[\frac{0.4}{(1+0.06)^{-9}} + \frac{0.3}{(1+0.06)^{-8}} + \frac{0.3}{(1+0.06)^{-7}} \right]$$

$$= 8,025,175 \ US\$$$

while in alternative B:

$$PW_{B,10} = 5,000,000 \times \left[\frac{0.6}{(1+0.06)^{-9}} + \frac{0.4}{(1+0.06)^{0}} \right] = 7,068,437 \ US\$$$

The annual repayments starting from this moment will be 0.0872 × 8,025,175 = 699,795 US\$ in alternative A, and 0.0872 × 7,068,437 = 616,368 US\$ for alternative B. These are to be paid for a period of 20 years.

Problem 4-2

Calculate the most economical diameter of the transmission line that transports an average flow Q = 400 m³/h. The price of energy can be assumed at 0.15 US\$ per kWh and the average pumping efficiency is 65 %. The cost of the pipe laying in US\$/m length can be determined from the linear formula 1200 × D where D is the pipe diameter expressed in metres; the friction factor of the pipe can be assumed at λ = 0.02.

The investment is going to be repaid from a 20-year loan with an interest rate of 8 %. What will the annual repayments be if the total length of the pipe is 1 km?

Answer:

From Equation 4.9, for a = 1200:

$$D = 0.05\sqrt{Q}\sqrt[6]{\frac{e}{\eta a_{n/r} a}} = 0.05\sqrt{400}\sqrt[6]{\frac{0.15}{0.65 \times 0.1019 \times 1200}} = 0.352 \ m \approx 350 \ mm$$

The annuity calculated according to the conditions of the loan will be:

$$a_{20/8} = \frac{0.08 \times (1+0.08)^{20}}{(1+0.08)^{20} - 1} = 0.1019$$

If the pipe length is 1 km, the total investment cost can be estimated at 1200 × 0.35 × 1000 = 420,000 US$, which results in annual instalments of 0.1019 × 420,000 = 42,798 US$

Problem 4-3

For the same pipe diameter and length from Problem 4-2, calculate the annual loss of energy due to friction and its total cost.

Answer:

For pipe D = 350 mm, L = 1000 m and λ = 0.02, the friction loss from the Darcy-Weisbach Equation for flow Q = 400 m³/h becomes:

$$\Delta H = \frac{\lambda L}{12.1 D^5} Q^2 = \frac{0.02 \times 1000}{12.1 \times 0.35^5}\left(\frac{400}{3600}\right)^2 = 3.89 \ mwc$$

The energy wasted on the friction loss on an annual basis will be calculated as:

$$E = \frac{\rho g Q \Delta H}{1000 \times \eta} T = \frac{1000 \times 9.81 \times 400 \times 3.89}{1000 \times 0.65 \times 3600} \times 24 \times 365 = 57,144 \ kWh$$

and its annual cost will be EC = 57,144 × 0.15 = 8572 US$. This calculation has no practical meaning, as the loss of energy due to pipe friction is unavoidable. This loss can however be reduced by increasing the pipe diameter.

Self-study: Spreadsheet Lesson 1-7 (Appendix 7)

4.2   Hydraulic design

The hydraulic design of water transport and distribution systems requires thorough calculations due to the significant impact of each component on the overall operation. Opting for a larger diameter, reservoir volume or pump unit will always offer more safety in supply but implies a substantial increase in investment costs. This reserve capacity can only be justified by estimating the potential risks of irregular situations; otherwise the distribution system will become in part a dead asset causing considerable maintenance problems.

4.2.1   Design criteria

Hydraulic design primarily deals with pressures and hydraulic gradients. In addition, the flow velocities and the pressure and flow fluctuations are also relevant design factors.

The pressure criterion is usually formulated as the minimum/maximum pressure required/allowed at the most critical point of the system. For instance, the USA guidelines of AWWA (2016-2) suggest a normal pressure range between 24.1 and 44.8 mwc (35 – 65 psi) requiring the use of pressure-reducing valves if the maximum pressure exceeds 55.2 mwc (80 psi). Also, the minimum pressure 'under all flow conditions' should be 13.8 mwc (20 psi) above the ground level.

Pressure requirements usually depend on company policy although they can also be standardised, i.e. prescribed by legislation. The starting point while setting the minimum pressure is the height of typical buildings present in the area, which in most urban areas consist of three to five floors. With pressure of 5-10 mwc remaining above the highest tap, this usually leads to a minimum pressure of 20-30 mwc above the street level. In the case of higher buildings, an internal boosting system is normally provided. In addition to this consideration, an important reason for keeping the pressure above a certain minimum can be firefighting.

Maximum pressure limitations are required to reduce the additional cost of pipe strengthening. Moreover, there is a direct relation between (high) pressure and leakages in the system. Generally speaking, pressures greater than 60-70 mwc should not be accepted. However, higher values of up to 100-120 mwc can be tolerated in hilly terrains where pressure zoning is not feasible. Pressure-reducing valves should be used in such cases. Table 4.2 shows pressure in the distribution systems of various cities around the world.

Table 4.2 Pressures in various cities around the world

City/country	Min. - max. (mwc)
Amsterdam / the Netherlands	±25
Vienna / Austria	40 – 120
Belgrade / Serbia	20 – 160
Brussels / Belgium	30 – 70
Chicago / USA	±30
Madrid / Spain	30 – 70
Moscow / Russia	30 – 75
Philadelphia / USA	20 – 80
Rio de Janeiro / Brazil	±25
Rome / Italy	±60
Sophia / Bulgaria	35 – 80

Source: Kujundžić, 1996.

The table shows a rather wide range of pressures in some cases, which is probably caused by the topography of the terrain. In contrast, in flat areas such as Amsterdam, it is easier to maintain lower and stable pressures.

In situations where water is scarce, the pressure is not thought of as a design parameter. For systems with roof tanks, a few metres of water column is sufficient to fill them. However, in some distribution areas, even that is difficult to achieve and the pressure has to be created individually (as shown in figures 1.15-1.17).

Besides maintaining the optimum range, pressure fluctuations are also important. Frequent variations in pressure during day and night can create operational problems, resulting in increased leakage and malfunctioning of water appliances. Reducing the pressure fluctuations in the system is therefore desirable.

The design criteria for hydraulic gradients depend on the adopted minimum and maximum pressures, the distance over which the water needs to be transported, local topographic circumstances and the size of the network, including possible future extensions. The following values can be accepted as a rule of thumb:

- 5-10 m/km, for small diameter pipes,
- 2-5 m/km, for medium diameter pipes, and
- 1-2 m/km, for large transportation pipes.

These values can be exceeded for the sake of improved conveyance, if the energy provision is reliable and the costs affordable.

Velocity range can also be adopted as a design criterion. Low velocities should be avoided for hygienic reasons, while overly high velocities cause exceptional head losses. Standard design velocities are:

- ± 1m/s in distribution systems,
- ± 1.5 m/s in transportation pipes, and
- 1-2 m/s in pumping stations.

4.2.2   Basic design principles

After the inventory of the present situation has been made, design goals have become clear and design parameters have been adopted, the next dilemma is in the choice of the distribution scheme and possible layout of the network. The following should be born in mind while thinking about the first alternatives:

1) *Water flows to any discharge point choosing the easiest path: either the shortest one or the one with the lowest resistance.*
2) *Optimal design from the hydraulic perspective results in a system that demands the least energy input for water conveyance.*

Translated into practical guidelines, this means:

- maximum utilisation of the existing topography (gravity),
- use of pipe diameters that generate low friction losses,
- as little pumping as necessary to guarantee the design pressures, and
- valve operation reduced to a minimum.

Yet, the hydraulic logic has its limitations. It should not be forgotten that the most effective way of reducing friction losses, by enlarging pipe diameters, consequently yields smaller velocities. Hence, it may appear difficult to optimise both pressures and velocities in the system. Furthermore, in systems where reliable and cheap energy is available, the cost calculations may show that the lower investment in pipes and reservoirs justifies the increased operational costs of pumping.

Hence, there are no rules of thumb regarding optimal pumping or ideal conveying capacity of the network. It is often true that more than one alternative can satisfy the main design parameters. Similar analysis to that shown in Figure 4.2 should therefore be conducted for a number of viable alternatives, calculating the total investment and operational costs per alternative (instead of the pipe diameter, as the figure shows). In any sensible alternative, higher investment costs will lead to lower operational costs; the optimal alternative will be the one where the sum of the investment and operational costs is at a minimum.

The first step in the design phase is to adopt an appropriate distribution scheme. Pumping is an obvious choice in flat areas and in situations where the supply point has a lower elevation than the distribution area. In all other cases, the system may be entirely, or at least partly, supplied by gravity; these situations were discussed in Chapter 3. The next step is in the choice of network configuration. Important considerations here are the spatial and temporal demand distribution and distances between the demand points, natural barriers, access for operation and maintenance, system reliability, possible future extensions, etc.

Water transport systems are commonly of a serial or branched type. Pipes will be laid in parallel if the consequences of possible failure affect large numbers of consumers, an industrial area or important public complex (e.g. an airport); two examples are shown in Figure 4.3. The layout of a water transport system often

Figure 4.3 Parallel water transportation pipes

results from the existing topography and locations of the urban settlements. An example of a branched transportation system is shown in Figure 4.4 for the Ramallah-El Bireh district of the Palestinian Authority. The system is located in a hilly area with elevations between 490-890 msl. It supplies approximately 200,000 consumers with an annual quantity of 9 million m³ (Abu Thaher, 1998).

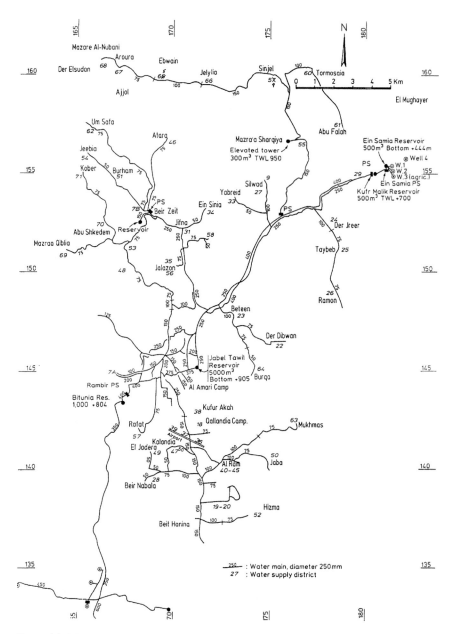

Figure 4.4 A branched water transport system in Palestine

Source: Abu-Thaher, 1998.

Creating loops is not typical for large transportation systems; such an approach is too expensive in many cases despite the shortcomings of a branched configuration. However, in smaller areas and with more favourable topographic conditions, this strategy may be feasible as it drastically improves the reliability of supply. An example from Figure 4.5 shows the regional system in the province of Flevoland in the Netherlands. The network of plastic pipes is laid in a sandy soil on a flat terrain; in 1996, it covered an area of approximately 230,000 consumers supplying an annual quantity of 15 million m³. Other water companies in the Netherlands have also created loops in their transport systems; compared to the examples mentioned in Chapter 1, these are comparatively smaller transportation systems and the 24-hour supply is a standard the Dutch consumers expect to be guaranteed for the price they pay for water.

Looped network configurations are common for urban distribution systems. How the layout should be developed depends on:

- the number and location of supply points,
- the demand distribution in the area, and
- future development of the area.

Firstly, the backbone of the system, made of large pipe diameters (secondary mains), has to be designed. If the network is supplied from one side, this can be of a branched structure. Characteristic of such a system is that the pipe diameters

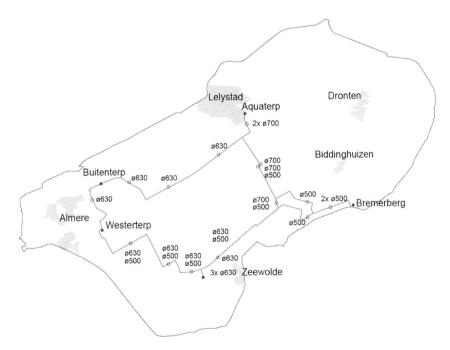

Figure 4.5 A looped water transport system

Source: Province of Flevoland, 1996.

will gradually reduce towards its end. A problem occurs if an alternative source, considered for future supply, is located on the opposite side of the network. Forming a loop (referred to as a 'ring') or a few major loops of the secondary mains is a better solution for this sort of problem, although more expensive.

The secondary mains often follow the routes of the main streets in the area, for the sake of easier access for maintenance and repair. A good starting point while selecting the main structure of the network is to examine the paths of the bulk flows, which can be determined for a known demand distribution in the area. If a network computer model is available, a preliminary test can be conducted by assuming uniform diameters in the system. The result of such simulation would show larger friction losses (velocities) in pipes carrying more water, indicating them as potential secondary mains.

An example of a distribution network showing a skeleton of the secondary mains is presented in Figure 4.6, for Zadar, a town in the coastal zone in Croatia. The gravity system supplies between 75,000 and 125,000 consumers (during the tourist season) with an average annual quantity of 8 million m³ (Gabrić, 1997).

In the second stage, the sizing of distribution pipes and analysis of the network hydraulic behaviour takes place. The support of a network computer model is fundamental here: the weak points in the system are easy to detect, and it is possible to anticipate the right type and size of the pumps and reservoirs needed in the system, as well as the additional pipe connections required. Alternatives that satisfy the main design criteria can further be tested on other aspects, such as operation under irregular situations, system maintenance, possible water quality deterioration, etc.

Transportation pipes that supply balancing reservoirs in the system are commonly designed for average flow conditions on the maximum consumption day. In distribution systems where 24-hour supply is a target, the network will be sized

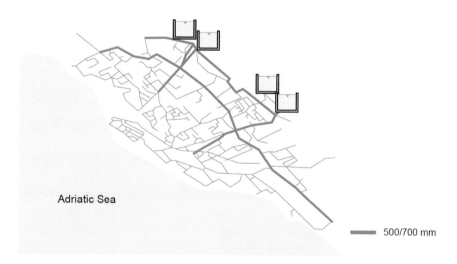

Adriatic Sea

500/700 mm

Figure 4.6 Layout of the distribution network in Zadar, Croatia

Source: Gabrić, 1997.

for the maximum consumption hour of the maximum consumption day. The ultimate buffer for safety is provided if, on top of that, a calamity situation is assumed
to take place at the same time: a fire or a failure of any of the system components.
As mentioned in Section 2.6 however, it may be more cost effective to let a limited
number of consumers 'enjoy' somewhat lower pressure or even an interruption
over a short period of time, rather than to specify pipes of a few per cent larger
diameter in considerable parts of the network in order to prevent a relatively rare
problem occurring. Such considerations constitute part of the reliability analysis
of the system, which is elaborated on further in Chapter 6.

Finally, the fire demand requirement is usually a dominant factor that influences the size of the pipes; in smaller pipe diameters it is actually a major con-
tributor to the peak demand compared to the regular demand. To avoid oversized
systems, the pipe diameters can be adopted based on the average hour demand
instead of the maximum hour demand on the maximum consumption day, in
addition to the fire demand. For instance, this is a common practice of many
water companies in the USA, which seems to offer a good balance between the
investment and any reliability concerns.

The points made in this and the remaining paragraphs of Section 4.2 are illustrated in a simplified design case of a medium-sized town, discussed in detail in
Appendix 3. The electronic materials available with this book can be used for a
better understanding of the exercise; the instructions for their use are also given
in Appendix 8. A cost comparison of the two developed design alternatives in the
exercise has been conducted according to the present worth method, which was
discussed in Section 4.1.2.

4.2.3   Storage design

While designing the storage volume, provision should be planned as in Figure 4.7.

The *demand balancing volume* depends on the demand variations. A 24-hour
demand balancing usually assumes constant (average) production feeding the
tank and variable demand supplied from it. Unless assumed to be constant, leakage should also be included in this balancing.

*Demand
balancing volume*

The calculation is based on the tank inflow/outflow balance for each hour.
Cumulative change in the tank volume (Equation 3.4) can be observed, and the
total balancing volume required is going to comprise the two extremes/provisions:

- the maximum accumulated volume stored when demand drops below average, and
- the maximum accumulated volume available when demand is above average.

The first provision needs to be available in the reservoir at the beginning of the
balancing interval i.e. at midnight. The normal trend will be that the balancing
volume first grows until the moment of the first hourly peak demand in the morning when the reservoir should be full (the second provision added to the first one).
The balancing volume will further be used for most of the day until the off-peak
period where it has been utilised fully and the first provision then needs to be
reinstated by the end of the 24-hour interval.

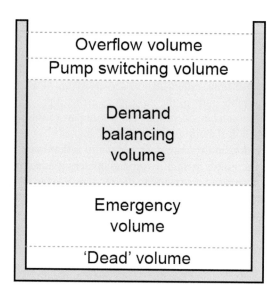

Figure 4.7 Volume requirements in a reservoir

The calculation procedure is illustrated in Table 4.3 for the diurnal demand pattern shown in Figure 4.8. The equal areas 1 and 2 in the figure are proportional to the balancing volume of the tank.

From the table: $V_{bal} = (2.27 + 2.21) \times 1489 = 6671\,m^3$, which is 18.6 % of the total daily demand of 35,737 m^3. The balancing volume (and therefore the total volume) is at its maximum at the end of hour 4. During the next hour, the diurnal peak factor becomes greater than 1 and the tank will start to lose its volume until the moment the peak factor drops below 1 again. This happens at the end of hour 19 when the balancing volume is completely exhausted. During the rest of the period the volume of the tank will be replenished back to the initial level at the beginning of the day. The required balancing volume at that moment is: $V_0 = 2.21 \times 1489 = 3291\,m^3$. Assuming a cross-section area A = 2500 m^2 and the minimum depth (incl. the reserve volume), $H_{min} = 2$ m, the level variation in the tank will be as shown in Figure 4.9.

Depending on the shape of the demand pattern, the balancing volume usually takes between 10 % and 30 % of the maximum day consumption. Generally smaller volumes are needed:

- for flat diurnal patterns,
- for diurnal patterns which fluctuate around the average flow, or
- if pumps in the system are operated to follow the demand pattern to some extent.

Examples of the three above cases are shown in figures 4.10-4.12, respectively.

Figure 4.12 shows the demand variation which is balanced by operating the pumps. Four equal units connected in parallel, each of them supplying 40% of the

Table 4.3 Example of the determination of the balancing volume

Hour	Q(m³/h)	pf	1-pf	Σ(1-pf)
1	579	0.4	0.61	0.61
2	523	0.4	0.65	1.26
3	644	0.4	0.57	1.83
4	835	0.6	0.44	**2.27**
5	1650	1.1	−0.1	2.16
6	1812	1.2	−0.2	1.94
7	1960	1.3	−0.3	1.63
8	1992	1.3	−0.3	1.29
9	1936	1.3	−0.3	0.99
10	1887	1.3	−0.3	0.72
11	1821	1.2	−0.2	0.5
12	1811	1.2	−0.2	0.28
13	1837	1.2	−0.2	0.05
14	1884	1.3	−0.3	−0.22
15	2011	1.4	−0.4	−0.57
16	2144	1.4	−0.4	−1.01
17	2187	1.5	−0.5	−1.48
18	2132	1.4	−0.4	−1.91
19	1932	1.3	−0.3	**−2.21**
20	1218	0.8	0.18	−2.02
21	898	0.6	0.4	−1.63
22	786	0.5	0.47	−1.16
23	657	0.4	0.56	−0.6
24	601	0.4	0.6	0
Avg.	1489	1		

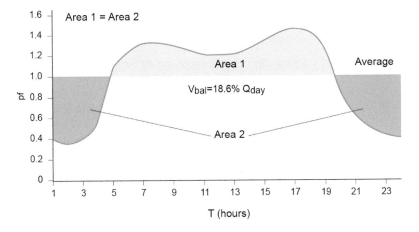

Figure 4.8 Relation between the demand pattern and balancing volume

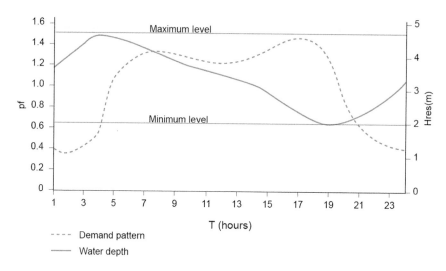

Figure 4.9 Relation between the demand pattern and reservoir water level variation

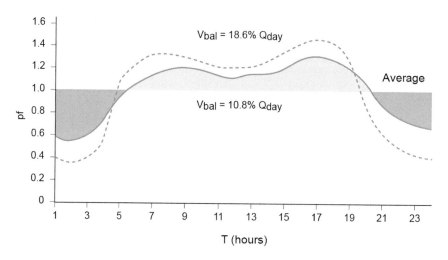

Figure 4.10 Balancing volume in the case of a flat diurnal pattern

average flow, are used to deliver the hourly demand. The first pump is in operation between hours 1 and 4, when the second unit is switched on. An hour later, the third unit starts operation and from hour 15 all four pumps are 'on'. This mode will continue until subsequent switching of three pumps takes place at hours 19, 20 and 21. The tank volume in this set-up is used for optimisation of the pumping schedule rather than to balance the entire demand variation. Without the tank, the fourth unit would have to operate for much longer, at least from hour 6, in order to guarantee the minimum pressures in the system; other units would have to change their operation, too.

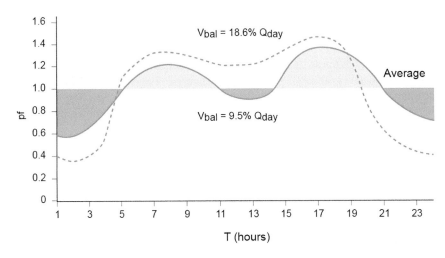

Figure 4.11 Balancing volume in the case of a fluctuating diurnal pattern

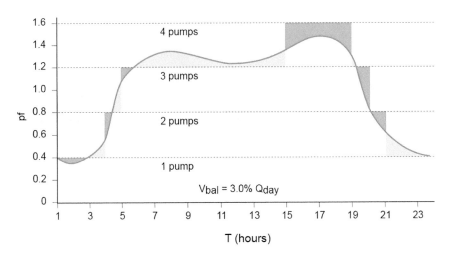

Figure 4.12 Balancing volume in the case of scheduled pumping

This example is typical for the operation of water towers. From the perspective of energy consumption, this is usually a more expensive solution than to pump the average flow continuously over 24 hours but the investments costs of the reservoir volume will be minimised. Hence, *the smaller the balancing volume is, the more pumping energy will be required.*

Applying a similar concept to that in Table 4.3, the required balancing volume can also be determined graphically. If the hourly water demand is expressed as a percentage of the total daily demand, this can be plotted as a cumulative water demand curve that will be compared to the corresponding cumulative supply curve.

In the example in Figure 4.13, for a constant supply over 24 hours, a straight line will represent the supply pattern. The required balancing volume equals the sum of the two extreme distances between the demand and supply curves (A-A' plus B-B'), which is about 28 % of the daily demand. The balancing volume available at the beginning of the day should equal the B-B' percentage. The tank will be full at the moment the A-A' percentage has been added to it and empty, i.e. at the reserve volume, when the B-B' deficit has been reached.

If the supply capacity is so high that the daily demand can be met with 12 hours of pumping a day, between 06:00 hours and 18:00 hours, the required storage is found to be C-C' plus D-D', which in this case is about 22% of the total peak day demand. However, if the same pumping takes place overnight or in intervals (in order to reduce the load on the electricity network i.e. save by pumping at a cheaper tariff), the required balancing volume will have to be much bigger. In the case of pumping between 18.00 hours and 06.00 hours, the balancing volume becomes C'-C''+ D'-D''≈ 76% of the daily demand. Hence, the *time period for intermittent pumping has implications for the size of the balancing volume*.

The volumes calculated as explained above (except for the water tower) are the volumes that balance the demand of the entire distribution area. These volumes can be shared between a few reservoirs, depending on their elevation and pumping regimes in the system. Optimal positioning and size of these reservoirs can be effectively determined with the support of a computer model. As Figure 4.14

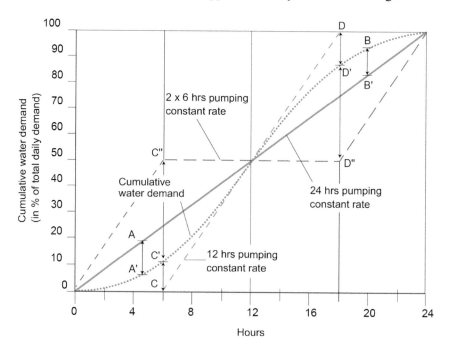

Figure 4.13 Example of graphical determination of the balancing volume

Source: IRC, 2002.

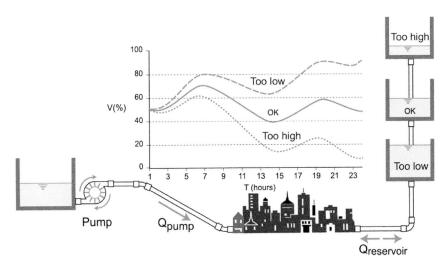

Figure 4.14 Relation between a tank's water level pattern and its altitude

shows, a correctly located reservoir more or less repeats the same water level pattern every 24 hours. A reservoir located too low soon becomes filled with water due to excessive pumping, while a reservoir located too high is going to dry out after some time due to insufficient pumping.

If better positioning of the tank is impossible, the pumping regime should be adjusted to correct the reservoir balancing. Such a measure will obviously have implications for pressure in the system.

Besides the balancing volume, other provisions in a reservoir include emergency volume, 'dead' volume, overflow volume and pump switching volume.

Emergency volume is exclusively used outside the regular supply conditions: *Emergency volume*

- during planned maintenance of the system,
- during a failure either in production facilities or somewhere else in the network, and
- for firefighting requirements.

How much water should be reserved depends primarily on how quickly the cause of interruption can be put under control. Each hour of average flow supply requires a volume equal to 4 %-5 % of the (maximum) daily demand. A few hours' reserve is reasonable, but more than that increases the costs of the tank, which also creates problems from water stagnation. Despite that, huge emergency volumes can be planned in large distribution areas. Special precautions then have to be taken to maintain the water quality in the tank (discussed further in Section 4.5.9).

'Dead' volume is never used. It is provided as a reserve that should prevent the *'Dead' volume* reservoir from staying dry. ±15cm of the depth is usually reserved for the 'dead' volume. More than that might be necessary if pumps that are supplied by the tank are located above the minimum water level. Certain provisions are required in

this situation in order to prevent under-pressure in the suction. The guideline suggested by the pump manufacturer KSB is presented in Figure 4.15. S_{min} from this figure equals $v^2/2g + 0.1m$, where v is the maximum velocity in the suction pipe.

Overflow volume

An *overflow volume* is provided as a protection against reservoir overflow. A few dozen cm of the depth can be allocated for this purpose. Within this range, the float valve should gradually close the inlet. For added safety, an outlet arrangement that brings the surplus water out of the system should be installed.

Pump switching volume

Pump switching volume is necessary if corresponding pumps operate automatically when there is a level variation in the tank. There is a potential danger if the pump switches on and off at the same depth: switching may happen too frequently (for example, more than once every 15 minutes) if the water level fluctuates around this critical depth. To prevent this, the switch-on and switch-off depths should be separated (see Figure 4.16). Depending on the volume of the tank, 15-20 cm of the total depth can be reserved for this purpose.

The hydraulics in the system may have an impact when selecting the inlet and outlet arrangements of a reservoir. Some examples are shown in Figure 4.17. The inflow from the top prevents backflow from the reservoir, while the outflow from the top usually serves as the second outlet (against overflow), or when certain minimum provision is to be guaranteed to specific users.

Self-study: Workshop problems 1.5.8 to 1.5.10 (Appendix 1)
Spreadsheet Lesson 8-10 (Appendix 7)

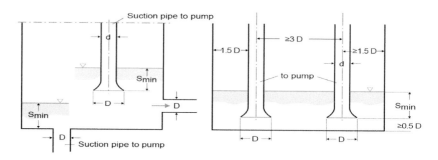

Figure 4.15 Minimum reservoir level where pumping is involved

Source: KSB, 1990.

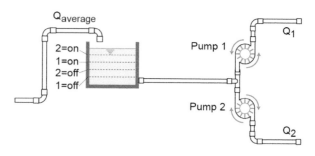

Figure 4.16 Pump switching levels in the reservoir

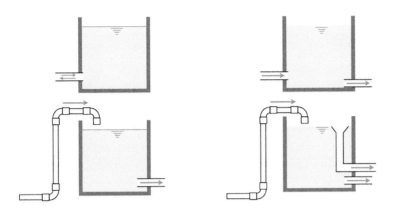

Figure 4.17 Reservoir inlet and outlet arrangements

4.2.4 Pumping station design

The capacity of a pumping station is usually divided between several units that are connected in parallel. A typical set-up consists of the elements shown in Figure 4.18.
   The role of particular components is as follows:

1) Valves are commonly installed at both the suction and pressure sides of the pump. These are used if the pump has to be dismantled and removed for overhaul or replacement. If necessary, a bypass can be used while this is being carried out. During regular operation of fixed speed pumps, the valve on the pressure side is sometimes throttled if the pumping head is too high.
2) A non-return valve (NRV) on the pressure side serves to prevent reverse flow.
3) An air valve on the pressure side is used to purge air out of the system.
4) The air vessel on the pressure side dampens the effects of transient flows that appear as soon as the pump is switched on or off, causing a pressure surge known as *water hammer*.   *Water hammer*
5) Measuring equipment: to register the pumping head, pressure gauges will be installed on both sides of the pump. A single flow meter is sufficient.
6) A cooling system can be installed to cool large pump motors.
7) Discharge pipes allow the emptying of the entire installation if needed for maintenance of the pipelines.

   The following main goals have to be achieved by the proper selection of pump units:

*   high efficiency, and
*   stable operation,

   Furthermore, the selected pumps should preferably have a similar number of working hours. This is easy to achieve if all the pumps are the same model (and impeller size), which allows their schedules to be rotated.

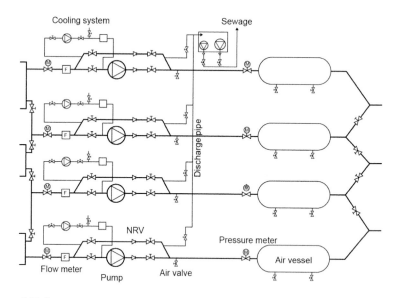

Figure 4.18 Pumping station layout

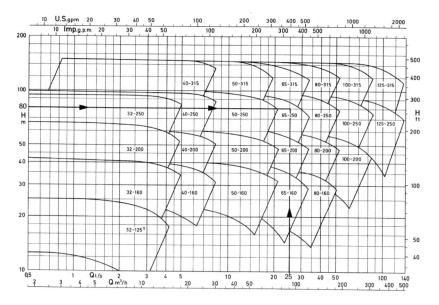

Figure 4.19 Operational regimes of pumps

Source: KSB, 1992.

In theory, the pumps that deliver duty head and duty flow are assumed to operate with optimal efficiency. These two parameters are used for the preliminary selection of pump units. The initial choice can be made from a diagram as shown in Figure 4.19. Such diagrams, showing operating ranges of various models, are commonly available from pump manufacturers.

It is possible to determine the impeller diameter, available net positive suction head and required pump power from the graphs related to a particular type (see Figure 4.20).

The pump power can also be calculated using Equation 3.73. In this case the values for the design head and flow assume the pump is operating under the maximum expected flow. A 10-15 % safety margin is normally added to the result; the first higher manufactured size will be adopted.

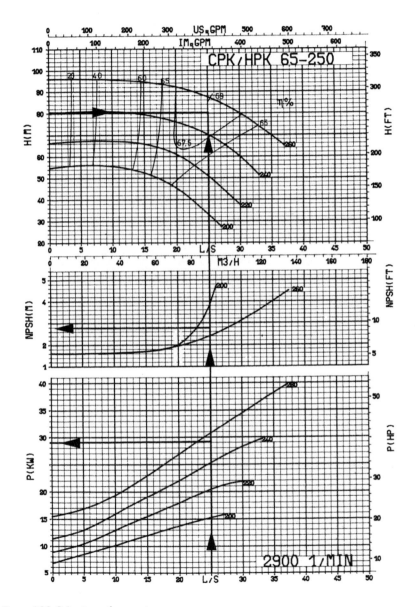

Figure 4.20 Selection of pump type

Source: KSB, 1990.

Rated power

For pumps driven by electrical motor, a transformer has to be sized. The capacity in kVA is calculated from the *rated power*:

$$N_{rated} = \frac{\sum_{i=1}^{n} N_{m,i} + N_{eq}}{\cos\theta}$$

(4.10)

where:
n = Number of pumps in operation under maximum supply conditions.
$N_{m,i}$ = Motor power per unit, calculated by Equation 3.74 in kW.
N_{eq} = Provision for other equipment in the pumping station: light, welding corner, etc.

Power factor

$\cos\theta$ = The *power factor,* which takes a value of between 0.7-0.8.

If a diesel generator is to be provided in the pumping station, its size will be designed to cover an electricity failure assumed to take place during the maximum supply conditions. With efficiency η_d, the generator power can be calculated from the following formula:

$$N_d = \frac{\sum_{i=1}^{n} N_{p,i} + N_{eq}}{\eta_d}$$

(4.11)

More energy is needed to start the pumps by diesel engine than by electricity. Therefore, to be on the safe side, the pump power of the largest unit, $N_{p,i}$, in Equation 4.11 is assumed to be doubled. Finally, the power needed to start the engine is:

$$N_{rated} = \frac{N_d}{\cos\theta}$$

(4.12)

Cavitation

The net positive suction head (NPSH) is the parameter used for risk analysis of *cavitation.* This phenomenon occurs in situations when the pressure at the suction side of the pump drops below the *vapour pressure*[1]. As a result, fine vapour bubbles are formed indicating the water is boiling at room temperature. When the water moves towards the area of high pressure, i.e. to the area around the impeller, the bubbles suddenly collapse causing dynamic forces, ultimately resulting in pump erosion. The damage becomes visible after the pump has been in operation for some time, and causes a reduction in the pump capacity (the actual pumping curve shifts lower than the original one).

The available NPSH is determined based on the elevation difference between the pump impeller axis (i.e. the centreline) and the minimum water level at the suction side. Two possible layouts are shown in figures 4.21 and 4.22.

1 The vapour pressure is the pressure at which water starts to boil. This pressure is dependent on altitude and affects the boiling temperature. As is well known, at mean sea level and normal atmospheric pressure, water boils at 100 °C. At higher altitudes the atmospheric pressure becomes lower and water will start to boil at a lower temperature.

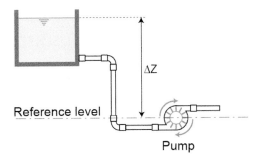

Figure 4.21 Pump unit located below the suction level in the reservoir

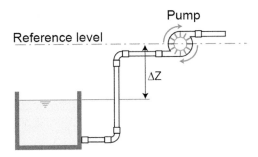

Figure 4.22 Pump unit located above the suction level in the reservoir

For the atmospheric pressure at the mean sea level of 10.33 mwc and standard range of water temperatures between 10 and 20 °C, the following simplified equation can be used at lower altitudes:

$$NPSH_{available} = 10 - \Delta H \pm \Delta Z - fs \qquad (4.13)$$

where:

ΔH = The total head loss along the pipe section between the reservoir and the pump, which includes all friction and minor losses between the suction and pump intake, including the entrance losses (mwc).

ΔZ = Elevation difference i.e. the static head between the pump axis and minimum suction level (m); the value of ΔZ becomes negative if the pump axis is located above the suction level.

fs = Factor of safety; a margin of 0.5-1.5 m is normally included.In the case of extreme altitudes, a more accurate calculation should take place by applying the following equation:

$$NPSH_{available} = h_{atm} - h_{vp} - \Delta H \pm \Delta Z - fs \qquad (4.14)$$

where:

h_{atm} = The atmospheric pressure (mwc); see Table 4.4.
h_{vp} = The vapour pressure (mwc); see Table 4.5.

Table 4.4 Atmospheric pressure	
Altitude (msl)	h_{atm} (mwc)
0	10.33
250	10.00
500	9.75
750	9.45
1000	9.20
1500	8.60
2000	8.10
3000	7.10

Table 4.5 Vapour pressure	
Water temperature (°C)	h_{vp} (mwc)
10	0.12
15	0.17
20	0.23
30	0.43
40	0.77
50	1.26
90	7.30
100	10.33

To prevent cavitation, the available NPSH has to be greater than the minimum NPSH required for the pump, which is read from diagrams such as the one in Figure 4.20. For the situations in Figure 4.22, the minimum required NPSH from the pump catalogue is commonly used to determine the maximum allowed distance between the minimum water level in the tank and the centreline of the pump impeller i.e. the axis, hence:

$$\Delta Z = h_{atm} - h_{vp} - \Delta H - NPSH_{required} - fs \tag{4.15}$$

To keep the head losses reasonably low, pipes in pumping stations are designed to maintain the optimal range of velocities; the recommended values are:

- feeder main v = 0.6-0.8 m/s
- suction pipe v = 0.8-1.2 m/s
- pressure pipe v = 1.5-2.0 m/s
- discharge header v = 1.2-1.7 m/s

A total head loss of a few metres of water column is possible in pumping stations. In addition to pipe friction, this is the result of lots of valves being installed, and bends created in order to 'pack' the pipes within a relatively small space. The obvious implication will be the reduction in the pump capacity. To keep this within acceptable limits, the total head loss should not exceed 1.0 mwc in the most extreme conditions, as a rule of thumb. To be able to limit the energy losses and meet the NPSH requirement, the reducers and enlargers will be constructed to reduce the pipe velocity; the recommended slopes are shown in Figure 4.23.

The critical head loss is calculated for the worst positioned pump unit, usually the last one in the pump arrangement, and under the maximum supply conditions. The friction and minor losses will be calculated according to the standard procedures explained in Section 3.2. Detailed tables for calculation of the minor losses are available in Appendix 5. Unless stated otherwise, all minor loss factors there are given for downstream flow velocity (i.e. after the obstruction).

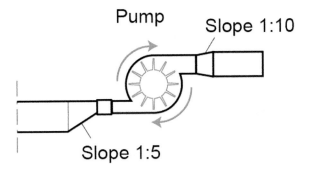

Figure 4.23 Pipe reducers

4.3   Computer models as design tools

Some 30 years ago, computer modelling of water distribution networks was carried out to only a very limited extent. The large-scale introduction of personal computers in the early nineties changed this situation entirely. The commercial programmes available on the market nowadays enable very accurate and quick calculations, even for networks consisting of thousands of pipes. These programmes are effective for use in the network design as well as for analyses of the system operation and planning of its maintenance.

The model of the Amsterdam network shown in Figure 4.24 was prepared using the older version (4.00) of 'InfoWorks WS' software, developed then by Wallingford Software Ltd of the UK. In 2005, such a programme would take just a few minutes to complete a 24-hour simulation of the distribution model consisting of nearly 40,000 pipes. Within such a short period of time, the entire system is calculated not once but 96 times, for every 15 minutes of the operation during a particular day! It is obviously ridiculous to compare the time used for manual hydraulic calculations against the time used for computer modelling, as they are on completely different scales, and in the meantime modern software has become a few dozen times faster.

Distribution network modelling programmes are all rather similar in concept, having the following common features:

- they run hydraulic simulations based on the principles explained in sections 3.5 and 3.6; the demand-driven calculations are executed as a default, while the pressure-driven demand calculations are used for irregular scenarios,
- they allow extended period hydraulic simulations,
- they possess an integrated module for water quality simulations,
- they can handle a network of virtually unlimited size,
- they can calculate complex configurations in a matter of minutes, and
- they have an excellent graphical interface for the presentation of results.

The main distinctions are in the specific format of input data used, as well as in the way the calculation results are processed and further optimised.

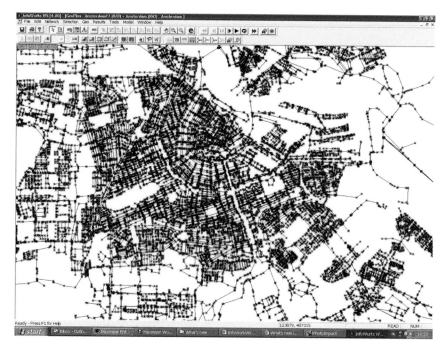

Figure 4.24 The Amsterdam distribution network model in 2005

Source: Wallingford Software Ltd.

Most of the algorithms on which the programme engine is based simulate the network operation by calculating a number of consecutive steady states. This number will be defined by selecting a uniform time interval between them (Δt discussed in Section 3.1), usually between a few minutes and one hour. Adjustment of the input data before each new calculation is done automatically, based on the results of the previous steady state calculation. Having the type and operation of the network components specified for the entire simulation period enables the programme to read the peak demands, recalculate storage volumes, switch the pumps on or off, etc. The final outputs of the simulation are the diagrams that describe the variations in pressure, surface level and flow in the system. The hydraulic results are further used as an input for the water quality simulation.

The entire modelling process consists of the following steps:

1) input data collection,
2) network schematisation,
3) model building,
4) model testing, and
5) problem analysis.

4.3.1   Input data collection

By possessing powerful computational tools nowadays, the focus in modelling has shifted from the calculation to the collection of reliable input data. High quality information concerning demand, system dimensions, materials and the maintenance level is crucial for accurate results; *quality of the input = quality of the output*. Well-conducted fieldwork data collection is therefore a very important initial step of the modelling procedure. The information to be investigated is listed below.

1   General

- Layout of the system – pipe routes and junctions; location of the main components.
- Topography - ground elevations in the area of the system; some specific natural barriers.
- Type of the system - distribution scheme: gravity, pumping, combined; role of each system component.
- Population - distribution and estimated growth.

2   Water demand

- Demand categories present in the system: domestic, industry, tourism, etc.
- Average consumption, patterns of variation: daily, weekly, and seasonal.
- Type of domestic water use: from direct supply, from roof tanks, etc.; average household size, habits with respect to water use.
- Demand forecasting.

3   Network layout

- Nodes (discharge points) - concerns predominantly the supply points of at least a few hundred consumers or major industry. Relevant for each point are:

  - location (X,Y) in the system,
  - ground elevation (Z), and
  - average consumption and dominant categories.

- Pipes - concerns predominantly the pipes $D \geq 80$-100 mm. Relevant for each pipe are:

  - length,
  - diameter (internal),
  - material and age, and
  - assessment of corrosion level (k, C_{hw} or N value if available).

- Service reservoirs - type (ground, elevated), capacity, minimum and maximum water level, shape (e.g. described through the volume-depth curve), inlet/outlet arrangement.
- Individual roof tanks (where applicable) - type and height of the tank, capacity, inflow/outflow arrangements, average number of users per

house connection, description of domestic installations (existence of direct supply in the ground floor).

- Pumping stations - number and type (variable, fixed speed) of pumps; duty head and flow and preferably the pump characteristics for each unit; age and condition of pumps.
- Others - description of appurtenances that may significantly influence the system operation (e.g. valves, measuring equipment, etc.)

4   System operation and monitoring

Important (preferably simultaneous) measurements for calibration of the model are:

- the pressure in a number of points covering the entire network,
- level variations in the service reservoirs and roof tanks (where applicable),
- pressures and flows in the pumping stations,
- the flows in a few main pipes in the network, and
- valve operation (where applicable).

In modern water distribution companies, much of the above information is directly accessible from online monitoring of the system. However, in many networks in the developing world, a large part of this information is missing or incomplete and the only real source available is the operator in the field. Even without lots of measuring equipment, some knowledge of the system is likely to exist in a descriptive form. For instance, in which period of the day a certain reservoir is empty/full, a certain pump on/off, a certain valve open/closed, a certain consumer with/without water or sufficient pressure, etc. Where there is a possibility of continuous measurements, similar days should be compared: the same day of the week in various seasons, or various days of the week in the same season.

5   System maintenance

Knowledge on the network maintenance, both preventive and reactive, helps to build a picture on the condition of assets for informed assessment in the absence of measured and recorded data. Typical points of attention include:

- Pipe-cleaning methods and frequency, for assessment of corrosion level or sediment accumulation affecting the decision on internal pipe roughness.
- Pipe repair and replacement frequency, for assessment of water losses.
- Pump maintenance and repair, for assessment of pumping curve deterioration resulting in a drop in efficiency and increased energy consumption.
- Valve maintenance, affecting the pressures and flows in the network.
- Practices of water conditioning in order to minimise negative impacts of its composition on the pipe material.

6   Water company

Analyses of water company practices are important because they indirectly contribute to the model upgrade for assessment of irregular demand scenarios and future demand growth. Some aspects to look at include:

- Organisation of monitoring and data collection, used to build the model and assess its accuracy.
- Measures in place to quickly react to various calamities, for assessment of service interruptions, water losses, emergency storage volumes, additional pumping capacity, etc.
- Procurement policies, in relation to the condition of assets.
- Standardisation and quality assurance applied in the renovation and expansion of the water supply system.

4.3.2 Network schematisation

The hydraulic calculation of looped networks is based on a system of equations with a complexity directly proportional to the size of the system. Thus, some schematisation (also called *skeletonisation*) of a network model may be preferred up to a level where the simulation run results will not be substantially affected while the computational time can be reduced. This was particularly important in former times, when computers were not extensively used and schematising the network could save several days (or even weeks) of calculation. In the meantime, this has become only a minor problem and current development of network modelling is taking the opposite direction; with the introduction of modern databases such as GIS, more and more detailed information is now included in analyses, specifically for monitoring of the network operation. As a result, the user is now confronted with a huge amount of data resulting in somewhat bulky computer models, not always easy to handle or understand properly. Schematisation therefore still remains an attractive option in situations where the design of the main system components is analysed, because:

Network skeletonisation

- it saves computer time,
- it allows model building in steps i.e. easier tracing of possible errors, and
- it provides a clearer picture about the global operation of the system.

Complex models of more than a few hundred pipes are not relevant for the design of distribution systems. Modelling of pipes under 100 mm significantly increases the model size without real benefit for the results. It also requires much more detailed information about the input, which is usually lacking. Thus, it is possible for a larger model to be less accurate than a small one.

The most common means of the network schematisation are:

- combination of a few demand points close to each other into one node,
- exclusion of a hydraulically irrelevant part of the network such as branches and dead ends at the borders of the system,
- neglecting small pipe diameters, and
- introduction of equivalent pipe diameters.

On the other hand, while applying the schematisation it is not permitted to:

- omit demand of excluded parts of the network, or
- neglect the impact of existing pumps, storage and valves.

The decision on how to treat the network during the process of schematisation is based on the hydraulic relevance of each of its components. It is sometimes desirable to include all the main pipes; in other situations pipes under a specific size can be excluded (e.g. below 200 mm). As a general guideline, small diameter pipes can be omitted:

- when laying perpendicular to the usual direction of flow,
- if conveying flows with extremely low velocities,
- when located in the vicinity of large diameter pipes, or
- when located far away from the supply points.

Whatever simplification technique is applied, the basic structure of the system should always remain intact, without removing the pipes that form the major loops. If the process has been properly conducted, the observed results of the schematised and full-size model for different supply conditions should deviate by not more than a few per cent.

An example of the network schematisation shows the simplified network in Hodaidah (already displayed in Figure 1.10) used for a computer model in a hydraulic study. Figure 4.25 shows the same network when pipes of 100, 200 and

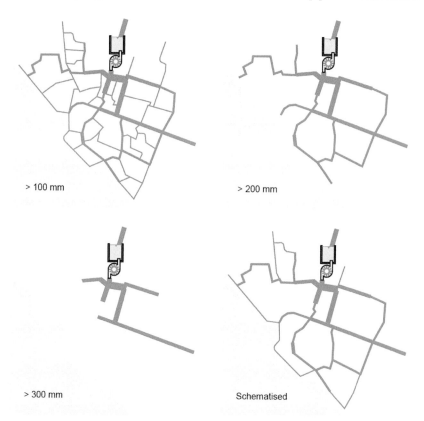

Figure 4.25 Example of network schematisation

300 mm diameters are removed successively, and finally shows the schematised layout that was used in the hydraulic calculations.

4.3.3 Model building

Computer programmes for network hydraulic modelling distinguish between two general groups of input data:

1. Junctions - describing sources, nodes and reservoirs (water towers),
2. Links - describing pipes, pumps and valves.

Although the way some components are modelled may differ from one to another software (for instance, pumps and valves need not necessarily be modelled as links and some programmes allow their modelling as junctions) the following input information is required in all cases:

*   Sources: identification, location and elevation of the water surface level.
*   Nodes: identification, location and elevation, average demand, and pattern of the demand variation.
*   Reservoirs: identification, position, top and bottom water levels, description of the shape (cross-section area, or the volume-depth diagram), initial water level at the beginning of the simulation, and the inlet/outlet arrangement.
*   Pipes: identification, length, diameter, description of roughness, and the minor loss factor.
*   Pumps: identification, description of pump characteristics, speed, and the operation mode.
*   Valves: identification, type of valve, diameter, head loss when fully open, and the operation mode.

For simulations of water quality, additional input information is required, such as: initial concentrations, patterns of variation at the source, decay coefficients, etc. Finally, a number of parameters that control the simulation run itself have to be specified in the input: duration of the simulation, time intervals, accuracy, preferred format of the output, etc.

Based on the above input, the raw results of the hydraulic simulation are flow patterns for links, and piezometric heads recalculated into pressures and water levels for junctions.

In addition, the water quality simulations offer the following patterns in each junction:

*   concentration of specified constituent,
*   water age, and
*   mixing of water from different sources.

In most cases, the input file format has to be strictly obeyed; this is the only code the programme can understand while reading the data. Programmes very rarely have exactly the same input format, so the risk of making errors during the

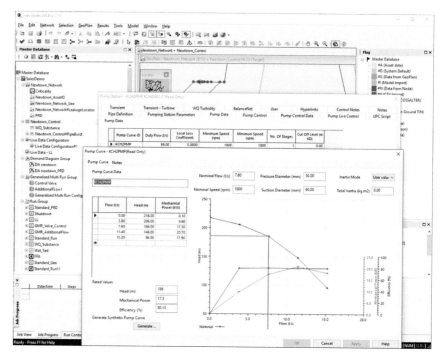

Figure 4.26 InfoWorks WS Pro (Ver. 2.5) – interactive data input

Source: Innovyze.

model building is very high. Modern programmes allow input in an interactive way, which reduces the risk of error caused by false definition of the network (see Figure 4.26). The disadvantage here is that the input data become scattered behind numerous menus and dialog boxes, often for each individual element of the network, which makes omission of some information fairly probable. Nevertheless, the testing of the network prior to the calculation is a standard part of any commercial software nowadays, and necessary feedback will be sent to the user depending on the library of error and warning messages.

Just as with networks in reality, *it is advisable to build the model in steps, gradually increasing the level of detail.* Starting immediately with the full size network, with all components included, will most likely yield problems during the model testing procedure. Nevertheless, this challenge is regularly encountered after using GIS as an input.

4.3.4   Nodal demands

A special aspect of the model building process is the determination of nodal demands. The problem arises from the need to survey numerous users spread all over the network and concentrate their demand into a limited number of pipe junctions in order to make the network presentation suitable for a computer model.

The starting point is the average demand calculation carried out with the equations discussed in Section 2.4. These equations yield the demand of a certain category or area, which has to be converted into demand at a point (pipe junctions). The next step is the conversion procedure which is based on the following assumptions:

- an even distribution of consumers, and
- the border between the supply areas of two nodes connected by a pipe is half way between them.

A unit consumption per metre of the pipe length can be established for each loop formed by m pipes:

$$q = \frac{Q_l}{\sum_{j=1}^{m} L_{j,l}} \tag{4.16}$$

Q_l is the average demand within loop l, and L_j the length of pipe j forming the loop. Each pipe supplies consumers within the loop by a flow equal to:

$$Q_{j,l} = q_1 \times L_{j,l} \tag{4.17}$$

and node i, connecting two pipes of loop l, will have the average consumption:

$$Q_{i,l} = \frac{Q_{j,l} + Q_{j+1,l}}{2} \tag{4.18}$$

One pipe often belongs to two neighbouring loops i.e. one node may supply the consumers from several loops. The final nodal consumption is determined after the above calculation has been completed for all loops in the system:

$$Q_i = \sum_{l=1}^{n} Q_{i,l} \tag{4.19}$$

n denotes the number of loops supplied by node i.

The procedure is illustrated in the example of two loops, shown in Figure 4.27. Average demands in areas A and B and lengths of the pipes are known data. Furthermore:

$$q_A = Q_A / (L_{1-2} + L_{4-5} + L_{1-4} + L_{2-5})$$
$$q_B = Q_B / (L_{2-3} + L_{5-6} + L_{2-5} + L_{3-6})$$

In loop A, the pipes supply:

$$Q_{1-2} = q_A \times L_{1-2}$$
$$Q_{4-5} = q_A \times L_{4-5}$$
$$Q_{1-4} = q_A \times L_{1-4}$$
$$Q_{2-5,A} = q_A \times L_{2-5}$$

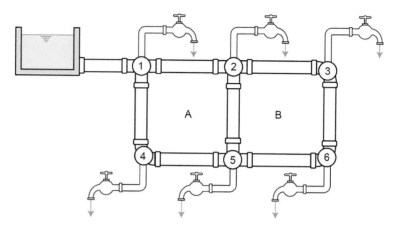

Figure 4.27 Example of nodal demands

and in loop B:

$$Q_{2-3} = q_B \times L_{2-3}$$
$$Q_{5-6} = q_B \times L_{5-6}$$
$$Q_{2-5,B} = q_B \times L_{2-5}$$
$$Q_{3-6} = q_B \times L_{3-6}$$

Pipe 2-5 appears twice in the calculation: once as a part of loop A, and the second time in loop B. Therefore, the nodal demands are:

$$Q_1 = (Q_{1-2} + Q_{1-4}) / 2$$
$$Q_2 = (Q_{1-2} + Q_{2-3} + Q_{2-5,A} + Q_{2-5,B}) / 2$$
$$Q_3 = (Q_{2-3} + Q_{3-6}) / 2$$
$$Q_4 = (Q_{1-4} + Q_{4-5}) / 2$$
$$Q_5 = (Q_{4-5} + Q_{5-6} + Q_{2-5,A} + Q_{2-5,B}) / 2$$
$$Q_6 = (Q_{5-6} + Q_{3-6}) / 2$$

For larger systems, a spreadsheet calculation is recommended. Alternatively, the same approach can be tried directly from the map by allocating a portion of the area of each loop to a corresponding node and determining the flow to be supplied from that node. From Figure 4.28:

$$Q_1 = Q_4 = 0.25 \times Q_A$$
$$Q_2 = Q_5 = 0.25 \times (Q_A + Q_B)$$
$$Q_3 = Q_6 = 0.25 \times Q_B$$

The above method is a simplification of reality, and good enough as an initial guess. Throughout the process of model calibration, the nodal demands calculated in this way need to be adjusted because of the impact of major consumers, large buildings, uneven leakage distribution, etc. Careful monitoring of the system as

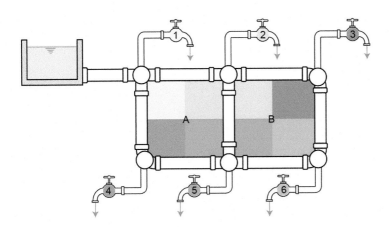

Figure 4.28 Nodal demand allocation

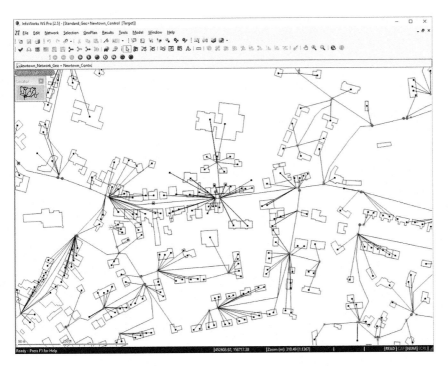

Figure 4.29 InfoWorks WS Pro (Ver. 2.5) – demand allocation
Source: Innovyze.

well as consistent billing records give crucial support in this situation. In more powerful programmes, this information can be directly allocated to the appropriate node, when the precise location of the service connection is known. An example is shown in Figure 4.29. In theory, every water meter can be transferred with its exact readings into a network model. However, although this can be an easy and automated process, it will result in bulky models that are complicated to analyse and maintain.

Self-study: Spreadsheet Lesson 8-11 (Appendix 7)

4.3.5   Model testing

Once the first simulation run has been completed, the immediate concern is whether the results match reality. In this phase, several runs have to be executed which must confirm that:

Model validation and calibration

- the model has a logical response to the altering of the input data; the simulation runs are in this case functioning in the *model validation*, and
- the model is behaving in relation to the real system; comparison of the calculation results with the hydraulic measurements is part of the *model calibration*.

There can be different reasons why the above conditions are not satisfied. The input file can be accepted by the programme as correct in syntax, but:

- Some input data were (badly) estimated, because the real values were not known.
- The network was transferred to the model with some typing errors or data was omitted.
- The format of the input file was incorrect but the error was not (clearly) defined in the error library: e.g. too high a calculation accuracy, insufficient maximum number of iterations, impossible operation mode specified, etc.
- The field measurements used for the model calibration were inaccurate.

Computer models cannot totally match a real situation; the results should always be judged based on the quality of input data and the measurements used for model calibration.

4.3.6   Problem analysis

With the correct execution of all the previous steps, the analysis of the problem is the final step of the modelling process and probably the shortest one. Some of the typical problems that can be solved by the help of a computer model are:

1. The selection of optimal pipe diameters for a given layout and demand scenario.
2. The selection of optimal models for pumps.
3. The selection of optimal position, elevation and volume of the reservoir(s).
4. The optimisation of the pump scheduling (to minimise energy consumption).
5. The optimisation of the reservoir operation (water depth variation)
6. The optimisation of the valve operation.
7. The simulation of fires.

8. The planning of pipe flushing in the system.
9. The analysis of failures of the main system components (risk assessment).
10. The analysis of water quality in the system (chlorine residuals, water age and mixing of water from various sources).

For many of these problems, the advantage of a quick calculation combined with proper analysis of the model response to the change in input data is that it will lead to correct conclusions on the network performance after a series of 'trial and error' simulations.

The new generation of computer programmes based on optimisation algorithms (genetic algorithms and neural networks) is trying to shorten the analysis even further. During the simulation process, the programme will attempt to satisfy a number of optimisation criteria set by the user. A typical example of an optimisation problem is the analysis of the least-cost pipe maintenance, in which the programme recommends the most economical measure from the list based on the network condition, the network performance and the unit cost of particular maintenance measures. Despite recent breakthroughs, these programmes are still in the continuous development stage and are yet to match the size of network and calculation times achievable by using traditional models.

There is a wide range of literature on the subject of water distribution network modelling. A very comprehensive overview of the methods and applications can be found in Walski *et al.* (Haestad Methods, 2003, reprint Bentley Institute Press, 2007).

4.4   Hydraulic design of small pipes

Computer models are rarely applied while designing systems in small residential areas and/or pipes of indoor installations; standardisation is a more popular approach for small (service) pipes rather than attempting to size them precisely. This ensures adequate system flows and pressures under ordinary supply conditions, avoiding detailed hydraulic analyses. Moreover, fire flows can often be a dominant component of design flows, overriding the peak flows caused by a relatively small number of consumers. Hence, adopting unique diameters makes maintenance easier and initially allows some buffer capacity for irregular situations. Some of the design methods are discussed in the following sections.

4.4.1   Equivalence Method

The Equivalence Method relates the design diameter of a pipe to the number of consumers that can be served by it. The equivalence table has to be prepared for each specific situation, based on the following input data:

- specific (average) consumption,
- a simultaneity diagram, and
- the design velocity or hydraulic gradient.

The calculation is normally carried out for a number of available (standard) diameters. A sample equivalence table is shown in Table 4.6 based on specific consumption of 170 l/c/d, the simultaneity curve from Figure 2.9 and velocity of 0.5 m/s.

The design capacity Q in the table is calculated based on the design velocity or the hydraulic gradient. For instance, in the case of pipe D = 60 mm:

$$Q_{60} = \frac{D^2\pi}{4}v = \frac{0.06^2 \times 3.14}{4} \times 0.5 \times 3600 \approx 5 \ m^3 \ / \ h$$

This flow has then to be compared with the peak supply condition, which is determined from the specific average consumption and the instantaneous peak factor for a corresponding number of consumers. In the above example:

$$Q_{60} = pf_{ins}n_iq_i = \frac{42 \times 17 \times 170}{1000 \times 24} \approx 5 \ m^3 \ / \ h$$

Correlation with the number of consumers may require a few trials before the results of the above two calculations match. Once the table is complete, a defined number of customers supplied from each pipe is directly converted to the design parameter. An illustration of the principle is shown in Figure 4.30 by using Table 4.6.

The Equivalence Method is used predominantly for the design of branched networks in localised distribution areas, from a few hundred up to a few thousand consumers. Simple looped systems can also be designed with this method after converting the grid into an imaginary branched type system. This can be done in several ways, most usually by 'cutting' the connections of the pipes expected to carry low flows (see Figure 4.31).

Practical difficulties in applying the Equivalence Method lie in the simultaneity diagrams, which are often difficult to generate because of the lack of reliable data. Creating an accurate instantaneous diagram requires careful monitoring of the network with a relatively long history of measurements. Alternatively, the diameters of the small pipes can be sized, based on the statistical analysis of the peak flows.

Table 4.6 Equivalence table (170 l/c/d, 0.5 m/s)

D(mm)	Q(m³/h)	pf_{ins}	Consumers
60	5	42	17
80	9	28	45
100	14	21	95
150	32	13	350
200	57	9	920
250	88	7	1900
300	127	5	3500

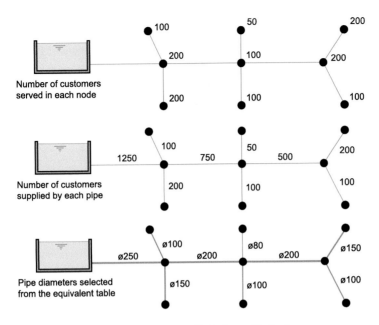

Number of customers
served in each node

Number of customers
supplied by each pipe

Pipe diameters selected
from the equivalent table

Figure 4.30 Example of pipe design using the Equivalence Method

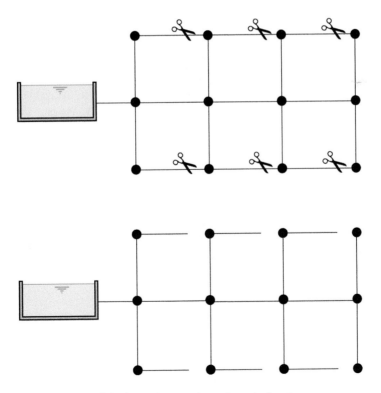

Figure 4.31 Conversion of the looped system into a branched system

4.4.2   Statistical methods

Statistical methods are predominantly used for the design of service connections and indoor pipes. They can also be an alternative to the Equivalence Method when the simultaneity diagram is not available.

In this approach the design diameter is directly related to the peak capacities, which were determined based on the locally established standards for indoor outlets.

In the Netherlands, the $q\sqrt{n}$-*method* offers an estimation of the peak instantaneous demand flow for a number of dwellings. The equation used is:

$$Q_{ins} = \sqrt{\sum_{i=1}^{r}\sum_{j=1}^{s} q_{i,j}^2}$$

(4.20)

where:
Q_{ins} = Peak instantaneous flow required by r dwellings.
r = Number of dwellings in the building.
s = Number of water outlets in one dwelling.
q = Standardised peak flow of each outlet.

To enable easy calculation, the maximum capacity of various types of outlets is expressed in so-called *tap units*. Equation 4.20 can then be transformed into:

$$Q_{ins} = q\sqrt{\sum_{i=1}^{r}\sum_{j=1}^{s}\left(\frac{q_{i,j}}{q}\right)^2} = q\sqrt{\sum_{i=1}^{r}\sum_{j=1}^{s} n_{i,j}}$$

(4.21)

where n is the number of tap units per outlet. This is a standard value, listed in Table 4.7 for q = 300 l/h.

A typical set-up of a family house in the Netherlands consists of one kitchen tap, one toilet with sink, and a bathroom with another toilet, sink, and bath tap, and a washing machine connection. The peak instantaneous flow in this case will be:

$$Q_{ins} = \frac{300}{3600}\times\sqrt{4+(0.25+0.25)+(0.25+1+4+4)} = 0.31 \ l/s$$

Table 4.7 Tap units, the Netherlands

Draw-off point	n_i
Toilet	0.25
Toilet sink	0.25
Bathroom sink	1
Shower	1
Bath tap	4
Washing machine	4
Kitchen tap	4

Source: VEWIN.

A similar approach is used in other European countries (Germany, Spain, Denmark, etc.).

4.5   Engineering design

Engineering design deals with non-hydraulic aspects of water transport and distribution systems. It is based on technical and financial grounds and tends to standardise the choice of components, materials, typical designs, and installation or construction procedures.

Decisions such as whether to opt for a limited choice of the most suitable pipe materials, a number of typical pipe diameters, pumping units supplied by the same manufacturer, and standardised valves, taps, bends, etc. that fit well with their dimensions, are of extreme importance as this obviously contributes to easier maintenance of the system and affects the value and volume of the stock. Water supply companies commonly adopt a material policy based on local conditions (costs, manufacturing, competitive supplies, delivery time, service, etc.) leading to standardisation in the entire supply area. In addition to this, standard designs for functional components of the water supply system, public taps, fire hydrants, house and garden connections, indoor installations, etc, need to be developed. In all of these activities, cooperation with the respective local industries normally provides a higher degree of self-reliance.

Finally, typical designs for pumping stations, elevated tanks, etc. can be proposed leading to a stage in which all these elements will be assembled into a kind

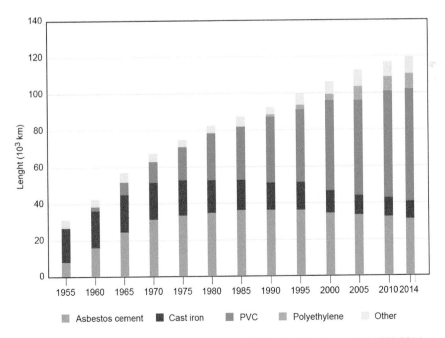

Figure 4.32 Development of pipe materials in the Netherlands in the period 1955-2014

Source: VEWIN.

of 'ready-made' project, allowing short preparation periods, accurate cost estimations, and uniformity. The feedback from experience in a number of such projects leads directly to improvements in the main design concept. The advantages for operational management and training purposes are also obvious.

For example, the experience in the use of pipe materials in the Netherlands shows interesting trends (Figure 4.32). The most widely used material in the first half of the 20th century was cast iron (CI). In the 1960s and 1970s, asbestos cement pipes (AC) were introduced on a major scale because they were corrosion-free, with thinner walls and therefore lighter than the cast iron pipes. In the 1980s and 1990s, the dominant pipe material became polyvinyl chloride (PVC) which was also corrosion-free but also more flexible, lighter and cheaper than asbestos cement yet still sufficiently strong owing to enhanced production technology, even in larger diameters.

As a result, the entire generation of CI pipes in the ground is nowadays some 80-100 years old, the AC pipes are mostly 40-60 years old, while the PVC pipes are usually less than 30-40 years old. Such a picture gives a clear idea as to when the moment for massive renovation of the networks is going to occur. Consequently, the replacement of the CI pipes has been largely carried out; it is expected that all the AC pipes will be removed from the ground within the next 20-30 years, whilst the replacement of the PVC pipes could be further postponed.

The current preference for flexible, plastic-based pipes is very logical for the Netherlands, given the predominantly unstable and corrosive soil conditions and predominantly flat topography where water pressures above 50-60 mwc are not common in the distribution networks. On the other hand, trends in other countries may be entirely different. In neighbouring Germany for instance, ductile iron pipes are more frequently used in many areas than PVC pipes, owing to different topographic and soil conditions. In any larger urban distribution network, a mixture of materials and pipe ages will be commonly encountered, often creating challenges in the management of these assets. Figures 4.33 and 4.34 show the situation in Belgrade, Serbia (BWS, 2015).

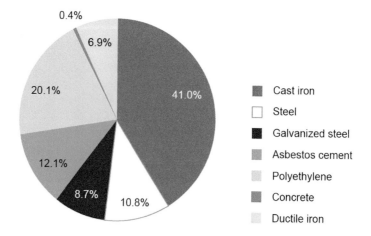

Figure 4.33 Pipe materials in the distribution network of Belgrade, Serbia
Source: BWS, 2015.

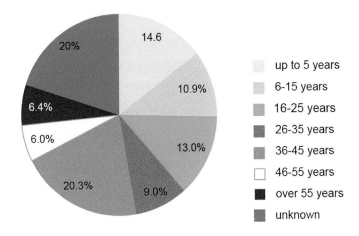

Figure 4.34 Pipe age in the distribution network of Belgrade, Serbia

Source: BWS, 2015.

4.5.1 Pipe materials

While deciding on the most suitable pipe materials, the following pipe properties should be taken into account besides the conveying capacity (Brandon, 1984):

- maximum and minimum depth of pipe cover,
- details on the backfill material,
- anticipated loading on ground surface,
- length and weight for handling and storage,
- resistance to corrosion and chemical action,
- permissible longitudinal and diametric deflection,
- pipe embedment and support conditions,
- ease of making repairs and future connections,
- ring stress to withstand heavy backfill loads without pipe deformation,
- in waterlogged ground, the weight of pipe in relation to the risk of flotation,
- pipe flexibility to be laid in a curved trench,
- pipe length with respect to the number of joints required,
- risk of damage from third parties, and
- deformations caused by extreme temperatures, uneven subsidence, vegetation roots, etc.

A careful choice helps to avoid the following problems in the operational management of transport and distribution systems:

- frequent interruptions in supply,
- increased water and energy losses,

- deterioration of water quality,
- shorter pipe lifetime, and
- expensive maintenance of the system.

Pipes used in water supply are made of various materials and, depending on their resistance to the backfill and shock loads, they can be categorised into three large groups:

- rigid: cast iron (CI), asbestos cement (AC), concrete,
- semi-rigid: ductile iron (DI), steel, and
- flexible: polyvinyl chloride (PVC), polyethylene (PE), glass-reinforced plastic (GRP).

Furthermore the pipe materials can be classified as metallic or non-metallic with obvious differences in the strength and corrosion properties being the most important. Despite the fact that the manufacturing process of plastic pipes has significantly improved in the last few decades, the traditional choice of metallic pipes in high pressure areas is still predominant, although this may be driven by other circumstances, such as freezing climate, for instance.

An ideal material that can be applied in all conditions therefore does not exist. Each of the pipe materials has a specific composition that determines its properties.

The main properties of the most commonly used materials according to the US experience (Smith et al., 2000) are compared in tables 4.8 and 4.9. Those materials are elaborated on further in this chapter.

Table 4.8 Properties of metallic materials in use in the USA

CHARACTERISTIC	CI	Lined DI	Steel	Galvanised steel (GS)	Copper tubes
Internal corrosion resistance	Poor	Good	Poor	Fair	Fair
External corrosion resistance	Fair	Moderate	Poor	Fair	Fair
Cost	Moderate	Moderate	Moderate	Moderate	Moderate
Specific weight	High	High	High	Moderate	High
Life expectancy	High	High	High	High	High
Primary use	T/D*	T/D	T/D	T/D	S
Tapping characteristics	Fair	Good	Good	Good	Good
Internal roughness	Moderate to high	Low	Moderate to high	Moderate to high	Low
Effect on water quality	High	Low	Moderate	Moderate	Moderate
Equipment needs	Moderate	High	Moderate	Moderate	Moderate to high
Ease of installation	Low to moderate	Low to moderate	Low to moderate	Low to moderate	Low Moderate
Joint water-tightness	Fair	Very good	Very good	Fair	Good
Pressure range (mwc)	NA	100-250	Varies	NA	Varies
Diameter range (mm)	NA	80-1600	100-3000	NA	<50
Ease of detection	Good	Good	Good	Good	Good

* T - transport, D - distribution, S - service connections
Source: Smith et al., 2000

Table 4.9 Properties of non-metallic materials in use in the USA

CHARACTERISTIC	AC	Reinforced concrete	PVC	PE	GRP
Material category	Concrete	Concrete	Plastic	Plastic	Composite
Internal corrosion resistance	Good	Good	Good	Good	Good
External corrosion resistance	Good	Good	Very good	Very good	Good
Cost	Low	Moderate	Low	Low	High
Specific weight	Moderate	Moderate	Low	Low	Low
Life expectancy	Moderate	High	Moderate	Moderate	High
Primary use	D*	T	D	S/D	Storage
Tapping characteristics	Fair	Fair	Poor	NA	NA
Internal roughness	Low to moderate	Low	Low	Low	Low
Effect on water quality	Low	Low to moderate	Moderate	Low	Low
Equipment needs	Moderate	High	Low	Low	Moderate
Ease of installation	Moderate	Low to moderate	Moderate to high	High	Low to moderate
Joint water-tightness	Good	Good	Good	Poor	NA
Pressure range (mwc)	70-140	Max. 160	Max. 160	Max. 140	NA
Diameter range (mm)	100-1100	300-4000	100-900	100-1600	NA
Ease of detection	Poor	Fair	Poor	Poor	Poor

T - transport, D - distribution, S - service connections

Source: Smith et al., 2000.

Cast iron is one of the oldest pipe materials used for the conveyance of water under pressure. However, the use of cast iron pipes has declined over the last few decades. Nevertheless, it is not uncommon to come across a cast iron pipe that is over 100 years old and still in good condition; considerable lengths exist in some networks even in developed countries, such as in the UK or USA. Two examples of old CI pipes shown in Figure 4.35 are from the Netherlands. These pipes have been well preserved even without much protection, owing to the continuous supply and absence of external corrosion, which was not the case with the pipes in Figure 4.36 originating from former Eastern Germany in the water distribution systems with intermittent supply and inadequate water composition. *Cast iron pipes*

Hence, the main disadvantage of CI pipes is generally low resistance to external and internal corrosion. This reduces the pipe capacity and iron release causes the water quality to deteriorate. Although the corrosion levels could stabilise after some time, the problems are usually solved by applying an internal cement lining and/or an external bituminous coating. The considerable weight and required wall thickness are additional disadvantages of these pipes.

Ductile iron is a material that has been in use for almost seventy years. It is an alloy of iron, carbon, silicon, with traces of manganese, sulphur and phosphorus. Unlike ordinary cast iron pipes made of grey iron where free carbon is present in flakes, the carbon in DI pipes is present in the form of discrete nodules (see Figure 4.37), which increases the material strength by 2-2.5 times. *Ductile iron pipes*

Figure 4.35 Old cast iron pipes in reasonably good condition

Figure 4.36 Heavily corroded cast iron pipes

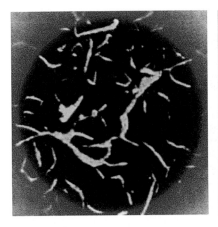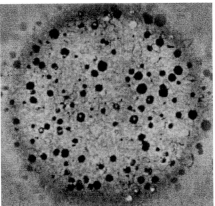

Figure 4.37 Texture of the grey iron and ductile iron materials
Source: Saint-Gobain.

DI pipes are strong, durable and smooth pipes, usually laid in mid-range diameters (100-600 mm) although they can be manufactured up to 1800 mm. They are suitable for almost all soil conditions if corrosion protected. Compared to cast iron pipes, DI pipes are lighter due to reduced wall thickness (see Figure 4.38) as well as being less susceptible to external loadings.

Despite their reduced weight, these pipes are still relatively heavy and handling with appropriate machinery is necessary even in smaller diameters; Figure 4.39 shows mid-range DI pipes stacked in the spare parts store. Also, in aggressive soil conditions, ductile iron may become susceptible to corrosion. As a minimum, bituminous paint should be used as an external protection. An internal cement lining is regularly applied to prevent corrosion resulting in turbidity and colour that may appear in water. The comparatively shorter length of these pipes results in a larger number of joints (as shown in Figure 4.40) increasing the concern for leakage.

Steel pipes are manufactured in two ways: by welding steel plates or stripes either longitudinally or in the form of a spiral (large diameters), or seamlessly from a steel billet (small diameters). These alternative manufacturing processes cover a wide range of diameters, wall thicknesses and fittings that can be applied to a variety of network layouts and working pressures. Typical diameters are between 100 and 1800 mm but more frequently these pipes are used for the transport of large water quantities. Steel pipes are the ultimate choice for piping in pumping stations (Figure 4.41). *Steel pipes*

Compared to iron or pre-stressed concrete pipes of the same internal diameter, steel pipes are stronger, more flexible and have thinner walls. Consequently, they are lighter and easier for handling and laying. Moreover, they are available in longer units, which reduces the total number of joints required (see Figure 4.42).

Secondly, repair of the pipes is relatively easy and can be conducted in a restricted space. Leaks are normally localised to joints or pinholes resulting in an acceptable loss of water. An additional safeguard in this respect is to weld the joints.

Figure 4.38 Cement-lined DI pipe with PE coating

Figure 4.39 Storage of DI pipes

Figure 4.40 Laying of DI pipes (large diameter – left-hand side, small diameter – right-hand side)

Figure 4.41 A spirally-welded steel pipe protected with anticorrosive paint

Figure 4.42 A cement-lined steel pipe with PE coating

Just as with other metal pipes, steel pipes are sensitive to corrosion. Hence, the internal and external protection must be perfect for both pipes and joints; an example of triple-layer coating is shown in Figure 4.43.

Galvanized steel (GS) is a zinc-coated pipe (both internally and externally) produced in smaller diameters with the aim of reducing corrosion. Nevertheless, water composition with a low pH may create acidic conditions resulting in the release of lead and cadmium possibly present as impurities in the zinc. They are not very expensive pipes but also not very strong. An example of a longitudinally-welded GS pipe is shown in Figure 4.44.

Copper pipes
A large number of service pipes and plumbing inside premises is made of *copper*. This material is popular being relatively cheap and reliable for large-scale implementation. Not used for distribution pipes, copper tubes rarely exceed 50 mm in diameter. The material is strong and durable, whilst at the same time sufficiently flexible to create any kind of bend, as shown in Figure 4.45. Low internal roughness helps to minimise the hydraulic losses resulting from long pipe lengths. Copper pipes are also easy to transport in straight pieces or coils, as shown in Figure 4.46.

Asbestos cement pipes
Asbestos cement pipes are rigid non-metallic pipes produced from a mixture of asbestos fibre, sand and cement (figures 4.47 and 4.48). The carcinogenic effect of asbestos-based materials used in water distribution has been studied carefully in the last couple of decades. Although not dangerous when in drinking water, the fibres can be harmful when inhaled. Therefore, the laying of new AC pipes has been prohibited by law in many countries, due to possible hazards during manufacturing, maintenance and disposal of these pipes.

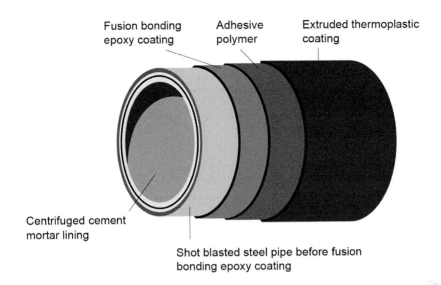

Figure 4.43 Steel pipe coating

Manufacturer: Europipe.

Figure 4.44 Longitudinally-welded galvanized steel (GS) pipe

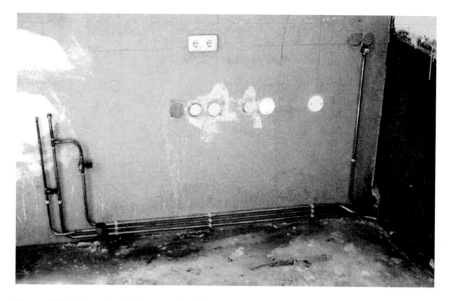

Figure 4.45 Indoor installations made of copper

Figure 4.46 Storage of copper pipes used in water distribution

The pipes that are still in the ground are typically of smaller diameters although larger diameters can also be found. The main advantages of AC pipes, compared to iron pipes, are:

- freedom from internal corrosion,
- generally better resistant to soil corrosion,
- a smooth inner surface,
- lighter weight, and
- lower production costs.

Figure 4.47 Asbestos cement (AC) pipes

Figure 4.48 Two AC pipes parallel to the coated steel pipe (in the middle)

The pipes can be drilled and tapped for service connections but are not as good as iron pipes in this respect; these locations are potential sources of leakages. Bursts and longitudinal deformations are also more likely and AC pipes in aggressive soils tend to corrode.

In the Netherlands, the manufacturing of AC pipes was stopped in 1993. The pipes still remaining in operation are handled with the utmost care; if they malfunction, the pipe will be replaced by a new one, usually made of PVC or PE.

Concrete pipes   *Concrete pipes* are rigid, cement-based pipes mainly used for sewerage. In drinking water supply, they will be more frequently laid for water transport than distribution. They are produced in diameters of between 250 mm and 1600 mm. They can occasionally be even larger, in which case they are almost always reinforced (Figure 4.49) and usually pre-stressed to withstand internal pressure and external loads. Pre-stressed concrete pipes can be reinforced in two ways: circumferentially with a steel cylinder, or both longitudinally and circumferentially with steel net (see Figure 4.50).

As a material, concrete is lighter than iron or steel but concrete pipes are generally heavier than the metal pipes of corresponding inner diameter, owing to a much thicker wall. For example, a pipe of D = 600 mm has a mass of some 500 kg per metre length, which creates difficulties during transportation and handling. This is why these pipes are produced in shorter lengths requiring more joints as a consequence (see Figure 4.51). Nevertheless, the advantage of this heavy weight

Figure 4.49 Old concrete pipes reinforced with steel net

Figure 4.50 Pre-stressed concrete

Figure 4.51 Ordinary concrete pipes used for non-pressurised flows

is that it limits any risk of movement, specifically in waterlogged ground. Furthermore, concrete pipes can carry heavy loads without damage or deformation and show good corrosion resistance; no special precaution is needed for pipe bedding or backfilling. Low internal roughness enables good hydraulic performance while conveying large water quantities. Last but not least, concrete pipes are comparatively cheap in large diameters.

Laying a route with concrete pipes requires more time than with metal pipes and involves heavy machinery. Operationally, a pipe repair also takes longer than with other types. If in the case of failure an alternative supply is not available, a larger part of the distribution area may be excluded from service. As with AC pipes, aggressive soil or water can cause corrosion which is normally restricted to the cement mortar but will attack the steel reinforcement once it is exposed to the environment; external coating is needed in such cases.

Figure 4.52 shows a 80-tonne pipe, D = 4000 mm, L = 7.5 m, made of pre-stressed concrete, used in the 'Great Manmade River' water transportation project in Libya.

Polyvinyl chloride pipes

Polyvinyl chloride pipes are flexible pipes widely used in a range of diameters up to 600 mm (see Figure 4.53). The properties of this thermoplastic material give the following advantages:

- excellent corrosion characteristics,
- lightweight,
- availability in long pieces,
- low production costs, and
- reduced installation costs.

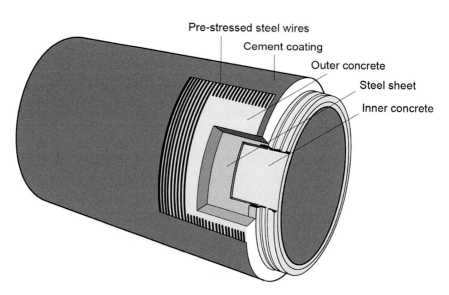

Pre-stressed steel wires
Cement coating
Outer concrete
Steel sheet
Inner concrete

Figure 4.52 Pre-stressed concrete pipe in the 'Great Manmade River' project in Libya
Source: The Management and Implementation Authority of the GMR Project, 1989.

Figure 4.53 PVC pipes in larger diameters

Table 4.10 Reduction in pressure rating of PVC pipes (Mays, 2000)

Service temperature (° C)	27	32	38	43	49	54	60
Percentage of original pressure rating	88	75	62	50	40	30	22

The disadvantages lie in the reduction of its impact strength in extremely low temperatures. An incidence of pipe bursts in wintertime has direct relation to the depth of the trench, which is adopted based on the comparison between the costs of excavation, pipe repair and estimated water losses. Furthermore, PVC pipes lose their tensile strength in extremely high temperatures. Table 4.10 shows the reduction of the pressure rating for PVC pipes depending on the operating temperature (Mays, 2000). As a result, careful handling, stacking and laying is crucial under extreme temperatures. Pipe support must be as uniform as possible, free from stones or other hard objects, and the filling material used must be well compacted.

Despite radical improvement in the manufacturing process, PVC pipes are uncommon in diameters above 600 mm. Larger diameters rarely withstand pressures much above 100 mwc and are therefore less suitable for high-pressure flows e.g. pressure pipes in pumping stations, or trunk mains. Moreover, a fracture can develop into a split along considerable lengths of the pipe, resulting in large water losses.

Pipe permeability

Another problem in operation occurs when the pipe is exposed to organic soil pollutants (oil, gasoline, etc) over long periods. Even in constantly pressurised pipes and without leakage, water quality may be affected after several months. This is a consequence of organic molecules passing through the pipe wall, which is known as *pipe permeability*. The result of it is taste and odour problems with considerable health hazards. In addition, the pipe material will soften, which weakens its structural strength. Laying of PVC pipes is therefore not advised in the vicinity of refineries or petrol stations.

PVC can be manufactured in different colours and particular colour codes can be adopted to recognise the pipes and the fluid they convey. In the Netherlands, drinking water PVC pipes are always cream-coloured (Figure 4.54), while the wastewater PVC pipes are grey and those transporting gas are yellow.

Polyethylene pipes

Another polymer used for pipe production is *polyethylene (PE)*. There are three phases in the development of the manufacturing technology resulting in the following types of PE pipes:

- low-density PE (LDPE), manufactured previously exclusively for service connections in diameters < 50 mm,
- medium-density PE (MDPE), with improved performance and for diameters up to 200 mm,
- high-density PE (HDPE), nowadays also manufactured in large diameters (exceptionally above 800 mm).

Figure 4.54 Typical PVC pipes used for distribution of drinking water in the Netherlands

Compared to PVC, PE pipes show the following enhanced characteristics:

- improved resistance to stress cracking,
- better performance under extreme temperatures,
- extreme flexibility, easier manufacturing of curves,
- good welding compatibility, fewer joints needed, and
- improved resistance to surge pressures.

Furthermore, handling of the pipes is easy. Smaller diameters (up to 100 mm) can be rolled in coils up to 150 m (Figure 4.55), while larger pipes are 10-12 m long (Figure 4.56).

Figure 4.55 Stock of coiled PE pipes

Figure 4.56 Large diameter PE pipes

One of the main disadvantages of PE compared to PVC is the higher price resulting from the thicker walls required for these pipes. Permeability of PE is also an issue. Maintenance problems mostly relate to jointing of the pipes. The welding technique is reliable but involves qualified personnel and electrical equipment. Also, special precautions have to be taken when using conventional mechanical fittings, due to the creeping of polyethylene. Water quality problems related to this material result from bio-film formation on the pipe wall.

Glass reinforced plastic pipes

Glass reinforced plastic (GRP) pipes are composed of three main components: fibreglass, resin and sand. The strength of the pipe is derived from bonding the fibreglass with resin. The purpose of adding sand is the increase in wall thickness that improves stiffness of the pipe, which makes its handling easier.

GRP pipes are commonly manufactured in larger diameters (see Figure 4.57) and thus used for transport of water. Compared with concrete and steel pipes of the same size, they are:

- lighter in weight,
- more flexible, and
- more resistant to corrosion.

In relation to the technical lifetime of pipes, GRP is still a relatively new material as it has only been in use since the 1980s. Therefore there is insufficient experience of its long-lasting performance to be able to draw firm conclusions. It is rather an expensive material with excellent corrosion characteristics, which

Figure 4.57 Glass-reinforced plastic (GRP) pipes

makes it predominantly suitable for industrial and chemical sites where intensive corrosive conditions may occur. Applications of this material frequently include the transport of wastewater. Other applications in drinking water supply include the lining of storage tanks.

Efforts have been made in the last couple of decades to judge pipe materials not only from the perspective of their performance in use but also on the effects they have on the environment either during manufacturing or in the post-exploitation phase. A typical example is PVC, which is considered to be a favourable pipe material in the Netherlands but which emits some carcinogenic pollutants during the manufacturing processes.

Although it is difficult to draw general conclusions due to the variety of criteria that have to be analysed, the assessment may be based on the following questions (Van den Hoven et al., 1988):

- How much of the natural reserves of raw materials that are used for pipe production are exhausted?
- How high is the energy input in production?
- What is the amount and structure of solid waste produced in the manufacturing processes?
- Are there hazardous substances released to the environment in the production phase?
- Can the pipe be re-used (re-cycled) after it has been exploited?

According to the authors, a comparison between pipes made of CI, steel, AC, PVC and GRP leads to the following conclusions:

- The environmental risks of exhaustion of natural reserves are acceptable in the case of the materials mentioned, with the exception of CI pipes when a zinc coating is applied. This effect can be reduced if a PE coating is used instead (Figure 4.58).
- Energy input in the pipe production is severe in the case of the CI and steel pipes (35 to 40×10^9 J per 100 m of the pipe length with a diameter of 100 mm). The figure for other pipe materials is 3 to 6 times lower, which is assumed to be reasonable.
- Emission of toxic substances, as shown in Table 4.11, can be observed during production.

CI, steel and PVC pipes score better than other materials in the post-exploitation period. The CI and steel can be completely re-used but coatings made of zinc or PE have to be renewed. PVC can be re-used 6-10 times resulting in slightly lower quality every new cycle. Re-cycling the GRP pipes saves around 30% of new raw material. AC remains dangerous because of the carcinogenic effect of AC fibres which is also in this phase.

As for the evaluation of their characteristics that is relevant for operation, any classification of pipe materials on an environmental basis is also difficult. AC is recognised as a potentially highly hazardous material but all the others also have adverse consequences in one or other phase. Pipe production is the most critical

Figure 4.58 Old scrapped CI pipes collected for recycling

Table 4.11 Emission of toxic substances as a result of pipe production

Pipe production	Emission into air	Pollution of water	Solid waste
CI, steel	CO_2 SO_2 NO_x	Sulphur Ammonium Phenol Cyanide Fluoride Lead Zinc Chloride	Heavy metals
AC	AC fibres		
PVC	SO_2 NO_x Chlorine Vinyl chloride		Vinyl Chloride
GRP	Heavy metals Acids	Hydrocarbon	

Source: Van den Hoven *et al.*, 1988.

aspect in this respect. Despite its limitations, this kind of analysis is obviously useful in pointing out critical manufacturing steps, which should be improved for minimising the environmental risks. In addition, it initiates research oriented towards possible new materials (e.g. glass-reinforced AC that might prevent the deficiencies of standard AC).

4.5.2 Joints

As with pipes, the joints that connect them can be classified as rigid, semi-rigid and flexible. For instance, flexibility in the routing of rigid pipes can be improved if flexible joints are used to connect them. On the other hand, some flexible pipes can also be welded to provide a safe watertight connection.

Standardisation of joints does not really exist. It is wise to limit the choice of joints to a few types; mixing different manufacturers and models may create stocking and repair difficulties. Poor jointing is often a major source of leakage. Hence, special attention should be paid to providing water tightness and protection from corrosion.

Welded joints are of a rigid type, suitable for steel and PE pipes. This is the cheapest joint for steel pipes of larger diameters, strong enough to carry high water pressure and longitudinal strain. These joints do not allow any pipe route deflection and a change in direction should be provided by proper fittings. Once the welding has been completed, the joint needs to be protected from external corrosion, which is done either by applying self-adhesive tape (Figure 4.59), or by melting a PE tape on top of the pipe external coating (Figure 4.60). Special care should be taken in the process of welding not to damage the internal cement lining of the pipe; manual repair of the lining may be considered in large diameter pipes. *Welded joints*

PE pipes are welded by thermal heat fusion methods. The most common process is called *butt fusion* (or *hot-plate fusion*); the ends of two consecutive pipes are cleaned, clamped, aligned and melted by a heated plate, and then fused at extreme pressure (well above the expected operational pressure; approximately 70-80 bar) for about 25-30 minutes (Figure 4.61 – left-hand side). This is a very convenient method because the jointing can be done before the pipes are laid, which enables the creation of long and flexible sections (Figure 4.62). As a result, narrow trenches can be used, which reduces the costs of excavation. Moreover, PE pipes of different density can be connected by this method. The mobile equipment for butt fusion is shown in Figure 4.63. *Butt fusion*

An alternative technique for welding PE pipes is called *electrofusion*. In this method a special type of dual socket coupling is used (known as *electrofusion coupling*, see Figure 4.61 – right-hand side), which contains parallel wires inside the perimeter; these convey the electric current used to melt both the pipe and the *Electrofusion*

Figure 4.59 Welding of steel pipes (protection with self-adhesive tape)

Figure 4.60 An alternative method of protecting the steel joint by melting the PE tape

Figure 4.61 Welding of PE pipes (left-hand side: butt fusion, right-hand side: electrofusion)

coupling. The process of welding has to be conducted by following the precise instructions of the manufacturer; an example of equipment and the cross section of a small diameter connection are shown in Figure 4.64; as can be seen, the wires stay in the coupling after the pipes have been connected. The instrument shown in the photo on the left-hand side is used to control the time, temperature and the

Figure 4.62 A large-diameter PE pipe section with joints created by the butt-fusion method

Figure 4.63 Butt-fusion welding equipment
Manufacturer: Ritmo.

Figure 4.64 Electrofusion equipment (left-hand side: the control box and coupling, right-hand side: the joint)

pressure, and to store the fusion parameters for documentation and inspection. The joints produced by the electrofusion need to cool down properly otherwise they may have reduced strength, which increases the risk of leaking.

Joints produced by butt fusion need longer straight lengths of the pipe, while electrofusion is a more expensive technique but easier to apply in a restricted space.

Flanged joints

Flanged joints are predominantly applied to rigid pipes, wherever a need for temporary disconnection occurs (pumping stations, cross-connection chambers, etc). They can be an integral part of the pipe, screwed on or welded to it. Connection bolts allow the work to be executed quickly while the rubber ring inserted between the flanges enables a watertight connection.

The flanged joints are mostly applied to metal pipes but they are also available with PE pipes, making the connection of these two materials possible. The example from Figure 4.65 (right-hand side) shows the jointing of PE and DI materials. Corrosion of the bolts may be a concern if the pipe is laid in aggressive soil.

Spigot and socket joints

Spigot and socket joints are flexible joints that allow a deflection between connected pipes. The traditional caulked lead spigot and socket joint was extensively implemented on cast iron pipes in the past. Nowadays, this type has been replaced by a variety of models that are also applicable for most other materials. The differences in the connection are mainly based on the filling of the space between the spigot and socket.

Sealing ring joints

Sealing ring joints (Figure 4.66, left-hand side) are made by a rubber gasket that is forced into a sealing position by the spigot and socket. The jointing is obtained by pushing one pipe into another (also known as 'push-in' joints). The water pressure adds to the tightening of the connection so that additional mechanical compression is not necessary. This type of connection is suitable for pipes made of pre-stressed concrete, DI, PVC and GRP. Deflection of the pipes is allowable from 1-5°, depending on the pipe material. The joint strength can be improved by welding of the pipes but this reduces the deflection. Another alternative is to coat the joint with polyethylene after the connection has been made.

Gland joints

Gland joints are connections where the rubber ring is forced by a rigid peripheral gland into a space between the spigot of one pipe and the socket of the next (figures 4.66 and 4.67, both right-hand side). This kind of connection has no

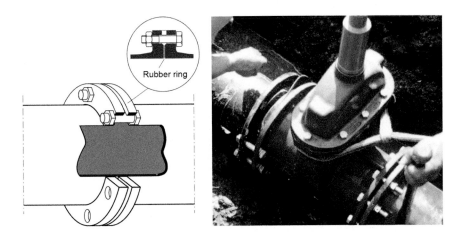

Figure 4.65 Flanged joints

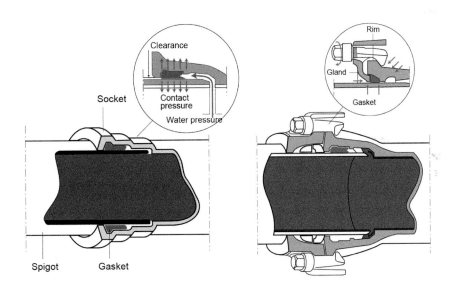

Figure 4.66 Spigot and socket (left-hand side) and gland joints (right-hand side)

significant longitudinal strength but allows deflections of 4-6º. It is commonly used for pipes made of CI and DI, but is also suitable when pipes made of steel and AC are connected to DI pipes.

PVC pipes can also be connected by dual socket push-in joints, which can be seen in Figure 4.68. Waternet, the water supply company for Amsterdam, guarantees additional tightness of the joint by inserting a flexible stick through the holes in the coupling, which surrounds the pipe preventing a joint burst.

Figure 4.67 Push-in and gland joints

Manufacturer: Saint-Gobain.

Figure 4.68 Dual socket joint for PVC pipes, with additional strengthening

Courtesy: Waternet, Amsterdam.

4.5.3 Fittings

Fittings are applied in situations that require a change in pipe diameter and/or material, pipeline direction, or when valves, water meters or hydrants have to be installed. A variety of designs exist. Several models made of ductile iron are shown in Figure 4.69.

It is normal to select pipes and joints from the same manufacturer although mixing of materials is also possible. For example: steel fittings can be used with pipes made of other common materials, PVC may be combined with high density PE, and DI and PE can also be jointed, etc. Figure 4.70 shows the combination of PVC and PE pipes connecting to a block of valves made of DI.

Figure 4.69 DI fittings

Manufacturer: Saint-Gobain.

Figure 4.70 Connection of PE and PVC pipes to a valve block made of DI

The mixture of different pipe materials and fittings is often caused by the routing of pipes in which some fittings require more space to deflect the pipes. For instance, a pipe bend made of PVC (Figure 4.71) will require more space than a similar one made of PE or DI.

The proper selection and installation of fittings is very important because they are, as with joints, very often a source of leakage. On most occasions, information about the construction, dimensions, application range, installation and maintenance is available in the manufacturer's catalogues. For instance, fittings for PE pipes can be welded by the butt fusion or the electrofusion methods but can also

Figure 4.71 A pipe bend made of PVC

Figure 4.72 Electrofusion fittings for PE pipes

be connected by a conventional welding technique called either *saddle fusion* or *socket fusion* in which both the outer surface of the pipe and the inner surface of the saddle/socket are simultaneously heated and kept pressed until the connection cools down. A collection of fittings that are installed by electrofusion is shown in Figure 4.72.

4.5.4   Valves

Valves in water distribution systems are distinguished by their principle of operation, the role in the system, and the manner of control. Generally, the valves fulfil three main tasks:

- flow and/or pressure regulation (flow control valves, pressure-reducing or pressure-sustaining valves),
- exclusion of parts of the network due to emergency or maintenance reasons (section valves), and
- protection of the reservoirs and pumps (e.g. float valves, non-return valves).

Operation of the closing element causes flow turbulence which creates a pressure loss. The magnitude of this loss is defined by the relationship between the valve position and the head-loss factor (a form of Equation 3.27). Manufacturers commonly supply a diagram known as the *valve characteristics* for each specific valve type and diameter. An example of this is shown in Figure 3.15.

Valve characteristics

Various valve constructions are based on the motion of the closing element, which can be linear (e.g. in the case of gate or needle valves), rotation (butterfly valves) or deformation (membrane/diaphragm valves). Gate and butterfly valves, as shown in Figure 4.73, are the most frequently used valves in water transport and distribution.

Gate valves function predominantly to isolate a pipe section; a valve block will be installed on an intersection between the pipes (Figure 4.74). Consequently, these valves normally operate in an open/closed position. Flow regulation is possible but is not common; the disk that is partly exposed to the flow may eventually loosen, causing leakage when it is in the closed position.

Gate valves

The hydraulic performance of gate valves is good, as the disc is fully lifted in the open position reducing minor losses to a minimum; this is also useful if the

Figure 4.73 Large gate valve (left-hand side) and butterfly valve with horizontal axis (right-hand side)

Figure 4.74. Blocks of gate valves.

pipe is cleaned mechanically. The disadvantage is, however, that the bonnet of the disc requires additional space around the valve.

As a prevention against surge pressures, the gate valves have to be continually open or closed for a long time, which makes them unsuitable in places where more frequent valve operation is required. It may sometimes take half an hour before a large gate valve is brought from one extreme position to the other. The process becomes even more difficult during the opening, as the thrust force acts only at one side of the disc. A bypass with a smaller valve is therefore recommended in the case of larger valve diameters, which is used to fill the empty section with water and to even up the pressures on the disc.

Butterfly valves   *Butterfly valves* have the disc permanently located in the pipe, rotating around a horizontal or vertical axis. When the valve is fully open, the disc will be positioned in line with the flow, creating an obstruction that increases the head loss compared to a fully open gate valve.

Butterfly valves are widely used in pumping stations as they are compact in size, and easier to operate and cheaper than comparable gate valves. They can also be applied in distribution networks, the main disadvantage being the obstruction created by the disc that makes mechanical cleaning of the pipe impossible. In both cases, the valve will be predominantly operated in an open/closed position but some degree of flow regulation is also possible (Figure 4.75). Nevertheless, using them for a high-pressure throttling over a longer period may damage the disc. Here as well, operating the valve operation too quickly is a potential source of surge pressures.

The number of section valves in any sizable distribution system can be huge, running into tens of thousands, with the vast majority of them not being frequently operated. An automatic device with an adjustable turning speed can be used in cases where many valves have to be operated within a short period of

Valve regeneration   time. Occasional turning of valves, known as *valve regeneration*, is a part of regular network maintenance in order to prevent clogging of the mechanism (figures 4.76 and 4.77).

Valves in water transport and distribution systems can also be distinguished according to their role in the system.

Figure 4.75 A partly opened butterfly valve with horizontal axis (the view from both sides)

Figure 4.76. Valve-turning machines

Manufacturer: Hydra-Stop.

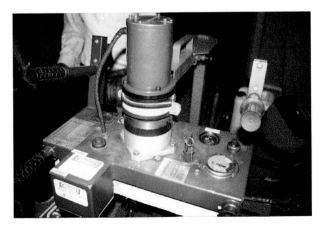

Figure 4.77 Control panel of the valve-turning machine

Non-return valve

Non-return valves (also known as check, retaining or reflux valves) are the valves that allow flow in one direction only. An opposite flow direction causes the valve to close and remain closed until the flow is re-established in its original direction. Hence, these valves operate in an on/off position, either fully closed or opened by the flow itself.

The non-return valves are installed in pumping stations (on discharge pipes) as backflow prevention; examples are shown in Figure 4.78. In distribution networks, the purpose of small non-return valves in the service connection is to prevent backflow contamination (Figure 4.79).

Float valve

Float valves are common devices in preventing reservoir overflows. They are automatically controlled by the surface water level in the reservoir. The principle of operation is shown in Figure 4.80.

In the figure, the closing element of a butterfly valve can be connected to a floating body, or a number of sensors at different elevations. Starting from a preset level, progressive throttling of the valve will occur as the water level rises,

Figure 4.78 A large non-return valve: cross section (left-hand side), and in operation (right-hand side)

Figure 4.79 Smaller non-return valves

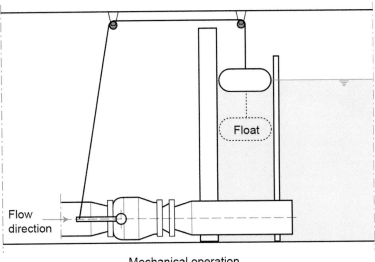

Mechanical operation

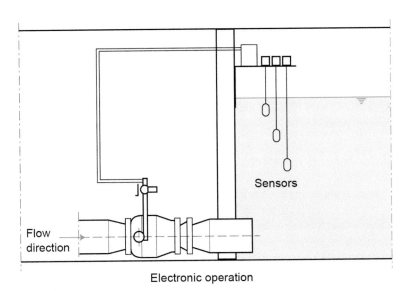

Electronic operation

Figure 4.80 Operation of the float valve

until the top level is reached. In this position, the valve will become fully closed. At the water level below the critical level, the valve remains fully opened.

Float valves can also be designed as diaphragm valves controlled by a pilot circuit connected to the float through a pilot valve; it is a small valve installed on the pilot tube operated on the pressure difference upstream and downstream of the main valve. Two models of these valves are shown in Figure 4.81.

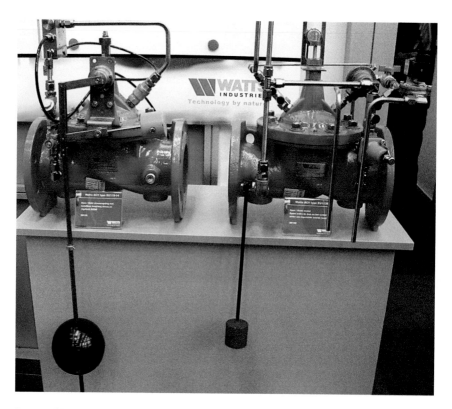

Figure 4.81 Float valves with diaphragm and pilot circuits
Manufacturer: Watts.

Pressure-reducing valves (PRV) are normally used to control the pressure in isolated parts of networks if it becomes too high. When the pressure upstream of the valve rises above the pre-set value, the valve will start closing until the downstream pressure is equal to the pre-set pressure. If the upstream pressure is below the pre-set value, the valve operates fully opened.

Pressure-reducing valves also operate as non-return valves: when the downstream pressure is higher than the upstream pressure, the valve is shut off. Consequently, these valves are equipped with upstream and downstream pressure gauges in order to maintain proper functioning.

As in the case of the valves shown in Figure 4.81, the principle of PRV operation is similar and involves a diaphragm valve with a pilot circuit; the small pilot valve on it is used to define the downstream pressure setting (Figure 4.82). PRVs can also be produced to have this setting variable i.e. remotely controlled based on the diurnal demand variation, providing more stable network pressures during 24 hours.

Pressure-sustaining valve

A *pressure-sustaining valve* (PSV) is in fact a pressure-reducing valve in reversed operation. In this case the isolated section of the network is upstream of the valve, where a particular minimum pressure needs to be guaranteed. The valve

Figure 4.82 Pressure-reducing valves (PRV)

Figure 4.83. PSV (left-hand side) and FCV with adjustable flow setting (right-hand side)
Manufacturer: Watts.

starts to close if the upstream pressure falls below the pre-set value. One model is shown in Figure 4.83 on the left-hand side.

A *flow-control valve* (FCV), as the name says, maintains a constant flow rate independent of the upstream pressure. With more sophisticated models, this flow setting can also be controlled remotely. An example of a FCV with an adjustable flow setting, which also prevents the upstream pressure from dropping below a particular minimum, is shown in Figure 4.83 on the right-hand side.

Flow-control valve

Due to their similar principle of operation, all the valves shown in figures 4.81-4.83 (float, PRV, PSV and FCV) can be categorised as *automatic-control valves* (ACV). These are produced in an even wider range of application than discussed in this section. The more detailed principle of operation and description of the particular valve components is available in the manufacturer's documentation (for example, for ACVs EU 100 Series made by Watts at www.wattsindustries.com).

Air valve

An *air valve* is a special type of valve that helps to release air from pipelines, which prevents reduction of the conveying capacity. Air accumulation can occur during the filling of the pipeline but also in normal operation. The valve consists of a float arrangement contained in a small chamber with an orifice vent. When water is present in the chamber, the pressure that initiates up thrust of the float closes the orifice. The appearance of air in the chamber depresses the water level uncovering the orifice. The air is expelled until normal water pressure in the chamber is established again.

Air valves are distinguished by the different size of orifice diameters, the number of chambers (single, dual) and the operating pressure. Examples of a single chamber valve are displayed in Figure 4.84 while the principle of operation of the dual chamber valve can be seen in Figure 4.85.

Figure 4.84. Single-chamber air valves.

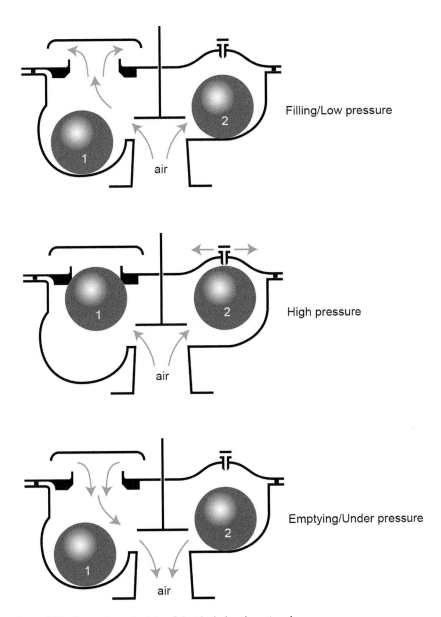

Figure 4.85 Operation principle of the dual-chamber air valve

The right-hand chamber functions as an ordinary air valve, whereas the left-hand (larger) chamber is used during low-pressure conditions: filling/emptying of the pipe or in any situation where negative pressure may occur.

Some confusion may exist when comparing air valves and *ball check valves* which look similar from the outside but are different in construction and operation. The latter are essentially non-return valves and the float inside the valve

Ball check valves

is used to block the reversed flow. Examples of ball check valves are shown in Figure 4.86.

Finally, valves can be operated manually or automatically. As partly explained, automatic operation is usually linked to a schedule (time-controlled valves), pressure or water level (float valves or pressure-controlled, pressure-reducing or sustaining valves), flow rate or flow direction (flow-controlled and non-return valves) somewhere in the system. A variety of mechanical, hydraulic and electronic equipment is nowadays involved in controlling valve operation, by sophisticated means that also allow remote operation. The electronic control devices are called *valve actuators*.

Valve actuators

The example in Figure 4.87 shows the butterfly valves with a vertical axis that can be controlled both manually and electronically. The operating mechanism

Figure 4.86. Ball check valves.

Figure 4.87. Manual and automatic butterfly valve operation.

should in both cases be equipped with an indicator of the position of the closing element. Valve operation in general assumes opening/closing at slow speed in order to prevent water hammer occurring in the pipe and potentially resulting in its severe damage.

4.5.5 Water meters

The purpose of metering in water distribution systems is twofold: it provides information about the hydraulic behaviour of the network, which is useful for the operation, maintenance and future design of the network extensions, as well as being the basis for water billing. In both cases accuracy is vital, so the quality and good maintenance of these devices are very important. Functioning of the water meters is based on three main principles:

1) pressure difference,
2) rotation, and
3) magnetic or ultrasonic waves.

Flow measurements based on the pressure difference are in fact applications of the Bernoulli Equation for two cross sections. This principle is used in the construction of 'Venturi' meters (Figure 4.88/up) and Orifice plates (Figure 4.88/down),   *'Venturi' meter*
Orifice plate

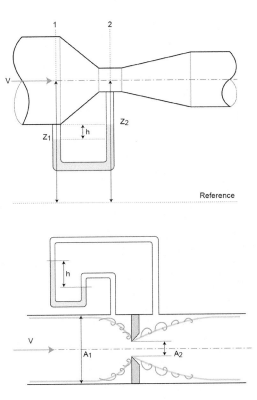

Figure 4.88 Hydraulic flow meters: a 'Venturi' meter (above) and an Orifice plate (below)

being mainly for the flow determination in trunk mains. These devices can also be found at major junctions of water transport systems, or within any water distribution system where large quantities are to be measured.

Both types are simple in construction and do not require any electronic equipment; a differential manometer is the only measuring device. The geometry of the cross sections and the minor loss coefficients based on the pipe contraction/shape of the orifice must be known for determination of the mean velocity; for further details, see Appendix 5.

Hydraulic flow meters present a degree of obstruction to the flow, generating hydraulic losses and limiting the pipe maintenance.

The flow meters that have no moving parts and do not present any physical obstruction are based on either magnetic field or ultrasonic wave measurements.

Magnetic flow meters use the principle of flow measurement based on Faraday's Law of Induction. In this, a certain voltage, U, is induced across a conductor (water in this case) moving through a perpendicular magnetic field of the strength, B, which is shown in Figure 4.89. As a result, a small electrical current will be generated proportional to the average velocity, which is calculated as:

Magnetic flow meter

$$v_{avg} = \frac{U}{kBD}$$

(4.22)

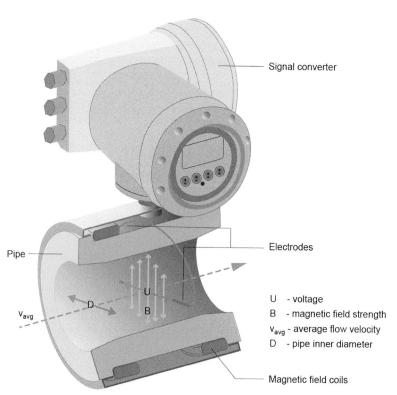

U - voltage
B - magnetic field strength
v_{avg} - average flow velocity
D - pipe inner diameter

Figure 4.89. Operation of magnetic meters

Adapted from: Hofmann, 2011.

where k is the device constant and D is the inner diameter. Sensor electrodes emit pulses to the signal converter, indicating the actual rate of flow, Q, which is further calculated as:

$$Q = v_{avg} \frac{D^2 \pi}{4} = U \frac{\pi D}{4kB} \qquad (4.23)$$

Magnetic meters are very accurate over a wide range of flows but are relatively sophisticated and therefore expensive devices for mass implementation. Although any upstream flow disturbances have a minimal effect on them, the manufacturers recommend installing them on a straight pipeline leaving a few meters of the straight run upstream of the meter and slightly less on the downstream side; these lengths are normally correlated to the pipe diameter. The measurements that can be bidirectional are virtually unaffected by the changes in water pressure and temperature, and do not inflict additional head loss. The range of applications commonly includes the measurements of bulk flows in pumping stations (Figure 4.90). The sensitive electronic components only require low power but transient flows, extreme ambient temperatures and/or wet (humid) conditions may restrict the application of these meters. They can also be vulnerable to lightning strikes, according to Van Zyl, (2011).

Ultrasonic flow meters make use of ultrasonic waves to sample the velocity profile within the pipe. The meters used for drinking water are commonly based on the *transit time principle*, which takes the speed of sound propagation in the water into account. In this approach, two sound transducers are installed along a short pipe distance, exchanging diagonal sound waves in opposite directions.

Ultrasonic flow meter

The difference between the sound frequencies of the two signals, which is proportional to the flow rate, will be registered because the sound travelling against the flow (from B to A, in Figure 4.91) will need more time to reach the transducer on the opposite side than the one travelling with the flow (from A to B); this is

Figure 4.90. Magnetic flow meters

Manufacturer: Krohne.

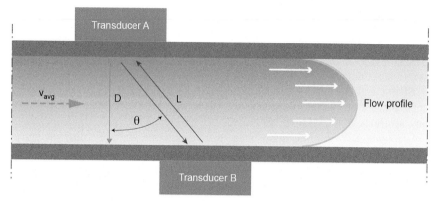

v_{avg} - average flow velocity
D   - pipe inner diameter
L   - distance between transducers
θ   - refraction angle

Figure 4.91 Transit time principle of ultrasonic flow measurement

easier to understand by applying the analogy of swimming diagonally from one river bank to another. Using further the notation in Figure 4.91:

$$L = \frac{D}{\cos\theta}$$

(4.24)

For c being the sound speed in the liquid:

$$c + v_{avg}\sin\theta = \frac{L}{t_{A\to B}} \quad ; \quad c - v_{avg}\sin\theta = \frac{L}{t_{B\to A}}$$

(4.25)

where t is the transit time of the sound pulse travelling between the transducers. Subtracting the equations in 4.25 and plugging in the one in 4.24 yields the relation between the average flow velocity and the transit times:

$$v_{avg} = \frac{D}{2\sin\theta\cos\theta}\left(\frac{1}{t_{A\to B}} - \frac{1}{t_{B\to A}}\right) = \frac{D}{\sin 2\theta}\frac{\Delta t}{t_{A\to B}t_{B\to A}}$$

(4.26)

Subsequently, the flow rate will be:

$$Q = \frac{kD^3\pi}{4\sin 2\theta}\frac{\Delta t}{t_{A\to B}t_{B\to A}}$$

(4.27)

where Δt is the time delay in the signal reception and k is the device coefficient obtained through calibration. In addition, a summation of the equations in 4.25 can also yield the formula for the sound speed, c:

$$c = \frac{L}{2}\left(\frac{1}{t_{A\to B}} + \frac{1}{t_{B\to A}}\right) = \frac{D}{2\cos\theta}\frac{t_{A\to B} + t_{B\to A}}{t_{A\to B}t_{B\to A}}$$

(4.28)

If the transducers are installed on the opposite side of the pipe, the wave exchange will be straight. However, when clamped on the same side of the pipe (Figure 4.92), they create a refraction of the emitted wave from the pipe wall prior to being received by the opposite transducer (Figure 4.93). This improves the accuracy of measuring by having longer signal delays, Δt, which is favourable in the case of smaller diameter pipes. In general, the axial beam injection is used for calibration purposes.

Clamp-on ultrasonic flow meters are less accurate but a cheaper solution than to use magnetic meters. Their main advantage is the ease of installation which allows mobile measurements on various sections of the network. The pipe diameter, material and wall thickness have to be taken into consideration while calibrating

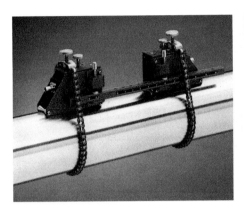

Figure 4.92 Clamp-on ultrasonic meters

Manufacturers: Controloton (left-hand side), Dynasonic (right-hand side).

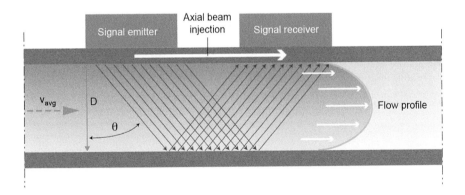

v_{avg} - average flow velocity
D   - pipe inner diameter
θ - refraction angle

Figure 4.93 Ultrasonic flow measurement by sound wave refraction

the measuring device. Ultrasonic flow meters can also be permanently installed on pipes (Figure 4.94), in which case they are closer in price and accuracy to magnetic flow meters. In either case, longer straight pipe sections should be selected upstream and downstream of the meter in order to minimize possible turbulences and formation of air bubbles which can reduce the measurement accuracy. Sediment formation in the pipe is also a concern because this reduces the inner diameter and can also disturb the signal.

Inferential meters   *Inferential meters* are mechanical meters used for flow measurements in small and medium size distribution pipes. These meters register the quantity of water passed by rotational speed of a vertical or horizontal rotor or vane, which is then transferred to a counter or register. Larger models are commonly produced in diameters of between 40 and 500 mm. Figure 4.95 shows two models used for

Figure 4.94 Ultrasonic meters for fixed installation

Source: (left-hand side: manufacturer: Teksan; right-hand side: a sophisticated model with additional acoustic leak detection feature, manufacturer: Kamstrup).

Figure 4.95 Inferential meters with a horizontal axis (left-hand side) and a vertical axis (right-hand side)

Figure 4.96 Household water meters used in the Netherlands: multi-jet, older type (left-hand side) and single-jet, newer type (right-hand side)

bulk measurements which both have an impeller with helical vanes (also known in practice as *Woltmann meters*).

Woltmann meters

Smaller inferential water meters are predominantly used for individual service connections in residential areas. They are produced with vertical vanes for pipe diameters of 15-40 mm and can be either the *single-jet* or *multi-jet* type (Figure 4.96). Newer single-jet models are more compact and lighter, yet sufficiently durable and accurate in operation.

Single-jet and multi-jet meters

Volumetric meters (also called *positive displacement meters*) are another type of mechanical meter used for billing purposes. The rotation of the moving element placed in a measuring chamber of a known size is converted to the volume of water registered on the display. The moving element can be a circular piston or a nutating disc. The principle of how the piston operates is shown in Figure 4.97.

Volumetric meters

Under normal operation (maximum working pressures around 100-120 mwc and ambient temperatures 0 to 40 °C), all small mechanical meters are fairly accurate and provide measurements within a ± 2% error margin. This error increases for very low flows, up to a lower limit of the operating range (typically ± 1 l/min); the meter cannot register flows below this limit. On the other hand, if the flow passing through is too high, the rotating element will be worn out quickly. A model with adequate nominal flow should therefore be selected. The manufacturers usually provide information on the operational flow range, working pressures, accuracy tolerance, etc.

Besides extreme working conditions, an additional concern with mechanical water meters is a high level of water hardness, causing clogging of the rotating elements after some time.

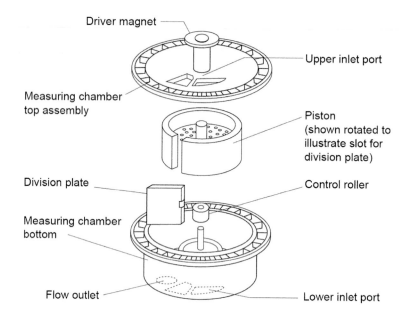

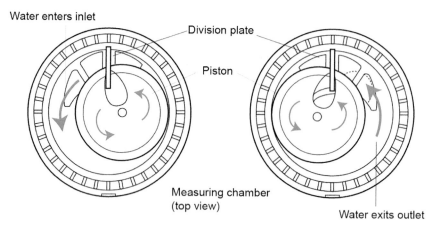

Figure 4.97 Principle for operation of the rotary piston
Adapted from: Coe, 1978.

Summary of the main properties of all the above-mentioned water meters is given in Table 4.12 (Van Zyl, 2011). D in the table indicates pipe diameter as a measure of the minimum lengths upstream and downstream of the meter.

The major use of water meters in any sizeable distribution system, specifically those used for billing purposes, often demands time-consuming work in collecting all the records (Figure 4.98).

Table 4.12 Main properties of water meters

Property	Meter type					
	Electro-magnetic flow meters	*Ultrasonic flow meters*	*Inferential meters*			*Volumetric meters*
			Woltmann	*Single-jet*	*Multi-jet*	*Rotary piston*
Common size (mm)	300 – 2000	400 – 4000	40 – 500	15 – 40	15 – 40	15 – 40
Sensitivity to velocity profile	Medium	High	High	Medium	Low	Insensitive
Sensitivity to water quality	Very low	Low	Low	Medium	Medium	High
Pressure loss	Very low	Very low	Medium	Low	Medium	High
Orientation	Almost any	Almost any	Almost any	Mainly horizontal	Horizontal	Any
Minimum straight length upstream	5 – 10D	10D	5D	0 – 5D	None	None
Minimum straight length downstream	3D	3D	3D	0 – 3D	None	None
Electricity driven	Yes	Yes	No	No	No	No

Adapted from: Van Zyl, 2011.

Figure 4.98 Manual reading of water meters (with the paper bill attached)

Figure 4.99 Remote reading of water meters

Despite substantial labour costs, manual reading of water meters is often subject to human error resulting from negligence or meter inaccuracy. However, the recent generation of household water meters are increasingly being read remotely, connecting them to loggers or sensors for wireless connection (Figure 4.99). Reading on the spot has also been possible without direct access to the water meter, which substantially increases the work efficiency due to better accuracy and also does not disturb the customers. As it is slowly becoming a practice in electricity and gas supply, modern water supply companies are gradually introducing *smart water meters* which are emitting continuous wireless signal about the water consumption without any need for intermediate meter readings. This approach enables more precise water balance between the demand and supply. One example of smart meter is shown in Figure 4.94 on the right-hand side.

A good practice in the Netherlands is that the customers declare their own consumption annually, either by post (Figure 4.100) or via the Internet, which is used as the basis to calculate the instalments they will pay in the following 12 months. The balance of the total sum paid and the water used is therefore made once a year. The room for abuse is very limited because the monthly instalments are usually automatically deducted from the customer's bank account and also the water supply company has a very good historical record of the water use at each address. Nevertheless, a check will be carried out in the event of suspicion, which happens rarely.

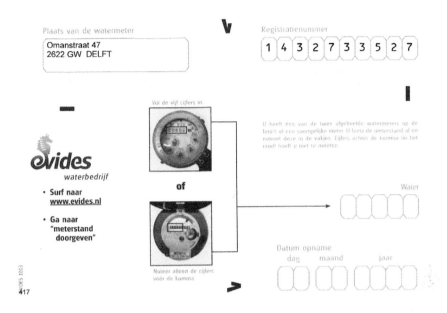

Figure 4.100 Customer reporting of meter reading in the Netherlands

4.5.6 Fire hydrants

Fire hydrants are generally distinguished as underground or ground installations (Figure 4.101). The underground installations are better protected from frost and traffic damage, but on the other hand they can be inaccessible when needed, for instance if they are covered by a parked vehicle. The exact position of a hydrant hidden by snow or ice can be detected in the Netherlands by a blue and red sign post indicating the exact distance and depth. The sign post distinguishes between valves (blue) and underground hydrants (red), as shown in Figure 4.101 (right-hand side).

The hydrants which are above ground are easy to detect; they are usually painted in bright colours: yellow, red, orange, etc., coding the various capacities of the hydrants. Nevertheless, many people consider them unaesthetic, and they can be damaged by cars or vandalised for illegal water use.

This type of hydrant is normally installed with a main valve, which keeps the barrel of the hydrant dry if not in operation. A small drain at the bottom allows the barrel to be emptied after the hydrant has been used (Figure 4.102, left-hand side). The advantages of such a set-up are that a potentially damaged hydrant is not going to leak. Moreover, illegal water use is impossible without access to the (underground) valve and finally, the freezing of water in the hydrant is prevented. In contrast, a wet barrel hydrant will be full with water all the time, allowing potential risks but also more prompt operation. Underground hydrants are also supplied with valves, primarily for maintenance purposes (Figure 4.103).

Figure 4.101 A ground-level fire hydrant (left-hand side), underground fire hydrant (centre) and sign post (right-hand side)

Figure 4.102. Dry barrel hydrants.

Figure 4.103 A block of gate valves with underground hydrant

Hydrants are usually located at intersections between streets in order to provide easy access from various directions (Figure 4.104). In a street, the distance between the two closest hydrants is around 100-200 m. Hydrants should not be placed too close to buildings as the vicinity of the fire or risk of the buildings collapsing might prevent their use. To avoid damage from traffic, they should also not be located too close to the road but if this is not avoidable, they should be protected from the initial impact (Figure 4.105).

The required capacity and pressure for hydrants varies from case to case and these are related to potential risks and consequences from fire. Generally, fire

Figure 4.104 Hydrant positioning at the intersection between streets

Figure 4.105 Hydrant protection against damage from parking vehicles

Figure 4.106 Water quality sampling using hydrants

requirements are within 30-50 m³/h, occasionally up to 100 m³/h, assuming minimum working pressures above 10-15 mwc. The pressure criterion is usually less of a concern, as the fire engine will normally be equipped with a booster pump. It is however logical to expect that the pressure in the distribution system will drop temporarily as a result of firefighting, affecting the neighbouring consumers to some extent. In the worst cases a vacuum can be created in the system, causing backflow contamination or surge pressures. This situation can be prevented if the hydrant is connected to a fire engine that has a (balancing) tank from where the water is drawn to extinguish the fire. Finally, the hydrants should be closed slowly in order to prevent surge pressures.

Apart from irregular situations, hydrants may be used for other purposes such as water quality sampling (Figure 4.106), cleaning of pipes, leakage control, flushing of streets, etc. Pipe networks with hydrants can also be separated from the drinking water distribution network.

4.5.7   Service connections

Service connections link users with the distribution system. The standard set-up consists of: the connection, a service pipe, outdoor and indoor stop valves, and a water meter. In newer installations, a non-return valve may be added as well.

A set-up with a stop valve used for flexible pipes is shown in Figure 4.107. Connection to the distribution pipe can be carried out on the top of the pipe, from the side, with or without saddle, etc. Service connections are commonly installed on distribution pipes in operation (i.e. under pressure). Direct access from the street to shut down the connection should always be provided for the purposes of maintenance, emergencies, or in the case of unpaid service. An example of a technique to install a service connection without a saddle on DI pipes is shown in Figure 4.108.

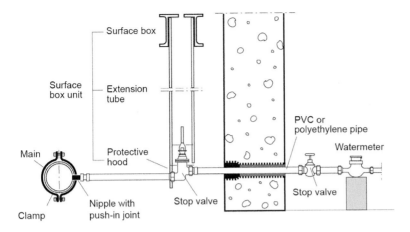

Figure 4.107 A household connection with a stop valve

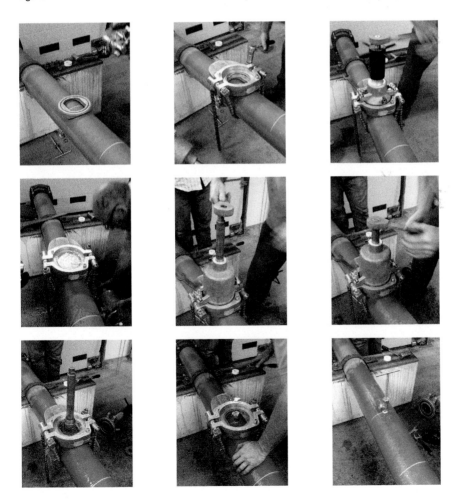

Figure 4.108 Service connection without a saddle on DI pipes

Courtesy: Waternet, Amsterdam.

Frequently used types of saddles for PVC and DI pipes in the Netherlands are shown in Figure 4.109. Usual diameters of the service pipes are 15-50 mm. Part of the pipe passing through the wall of the building will be sleeved by elastic and watertight material (plastic, impregnated rope, bitumen, etc). The connection of the service pipe to the metering set-up inside a house is shown in Figure 4.110.

Figure 4.109 PVC (left) and DI pipe (right) saddles commonly used for vertical service connections in the Netherlands

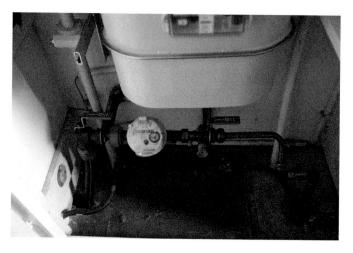

Figure 4.110 Indoor metering set-up connected to the PE service pipe (on the left)

This is usually the point where the responsibly of the water company to supply stops (after the meter), and the maintenance of the household installations and prevention of indoor water losses is done at the consumer's cost.

4.5.8   Indoor installations

Indoor systems comprise all the piping, tap points and appliances within dwellings. In higher buildings, a central pressure-boosting system is a compulsory requirement. The task of this installation is to bring pressure to the upper floors of the building. A tank can be constructed on the top of the building, in order to maintain the pressure but also as a reserve for fire protection. The example in Figure 4.111 shows a solution by which the pressure from the distribution pipe

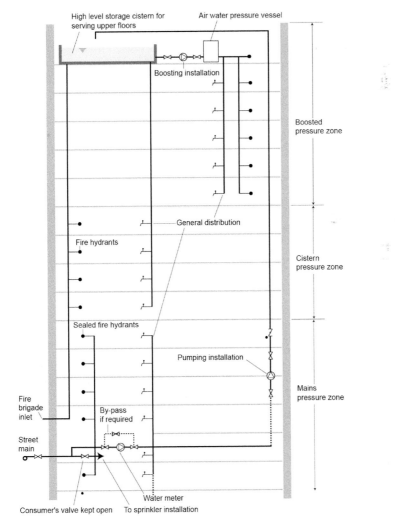

Figure 4.111 A high building with a roof tank

Source: Coe, 1978.

supplies only the lower floors, while the middle and the upper floors are supplied from the reservoir on the top. Two pumps are installed, one to fill the reservoir and the other to boost the pressure on the highest floors. Regarding household installations, there is a whole variety of configurations related to the habits of water use. An example of a European set-up is shown in Figure 4.112.

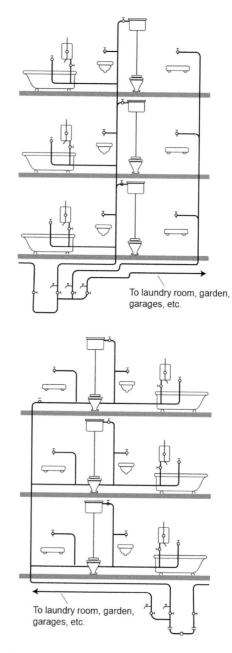

To laundry room, garden, garages, etc.

To laundry room, garden, garages, etc.

Figure 4.112 Indoor installations

Source: Coe, 1978.

4.5.9   Engineering design of storage and pumping stations

The engineering design of storage and pumping stations involves experts special-ised in construction, mechanical and electrical equipment. In addition to sufficient amounts of water, a well-designed storage should:

- maintain non-degraded water quality,
- provide reliable operation, and
- allow easy maintenance.

To reach the above goals, the following engineering aspects of storage construc-tion have to be considered (Williams and Culp, 1986):

- Water circulation through reservoirs should be provided by baffles, or by placing inlets and outlets on opposite sides of the reservoirs, with outlets near the bottom.
- Reservoirs should allow maintenance without loss of pressure in the distribu-tion system. If there is a single reservoir, it should be divided into compart-ments so that at least one part of the volume is available for use.
- Proper protection against contamination should be provided. All finished water storage structures should have suitable watertight roofs or covers to exclude surface water, birds, animals, insects and excessive dust. Suffi-cient ventilation should be enabled but protected against the same impacts (as shown in Figure 4.113). The drains should discharge into the ground sur-face with no direct connection to a sewer or storm drain.
- Water in the tank should be insulated against the impact of excessive tem-peratures (see Figure 4.114).
- Pipes connected to the tank should be equipped with valves installed outside the storage (except the overflow). Any pipe running through the roof or sidewall should be welded or properly gasketed in metal tanks or connected to built-in concrete castings with flanges. Corrosion protection should be introduced wherever necessary. The entry of silt through inlet pipes should be prevented.
- Access to the storage interior should be convenient for cleaning, mainte-nance and sampling. On the other hand, fencing, locks on access manholes and other necessary precautions should hinder vandalism and sabotage (Figure 4.115).
- Adequate measuring equipment should monitor the water levels in the tank.

The cross section of storage reservoirs is usually of rectangular or circular shape. An interesting construction was carried out in the case of the underground storage reservoirs in Munich (Germany) shown in Figure 4.116. The inlet struc-ture of this tank of 10,000 m³ is a perforated wall with openings of various diam-eters designed to create unequal flow distribution. The shape of the tank, which causes various retention times and mixing of the water, further amplifies this. More traditional inlet and outlet arrangements are shown in figures 4.117 and 4.118 (Obradović, 1992). In cases where the water is pumped from the reservoir, the suction pipe will be positioned below the bottom level of the tank in order to minimise its 'dead' volume and at the same time to guarantee the minimum suc-tion level for the pumps (Figure 4.118, left-hand side).

Figure 4.113 Protected air vents on an underground reservoir

Figure 4.114 Reservoir water heating to prevent freezing during winters in Ulaanbaatar, Mongolia

Figure 4.115 Restricted access (left-hand side) and monitoring of reservoirs water depth by float (right-hand side)

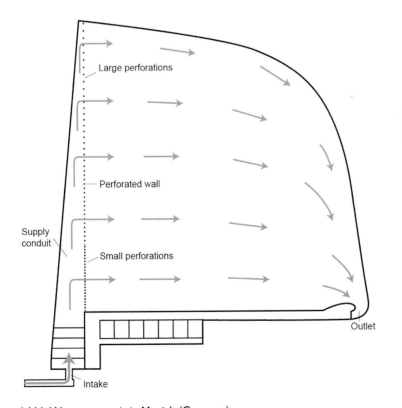

Figure 4.116 Water reservoir in Munich (Germany)

Inlet

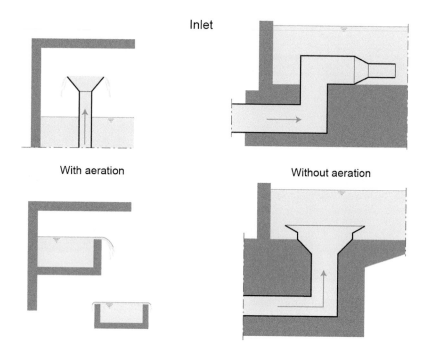

With aeration   Without aeration

Figure 4.117 Reservoir inlet arrangements
Source: Obradović, 1992.

Outlet

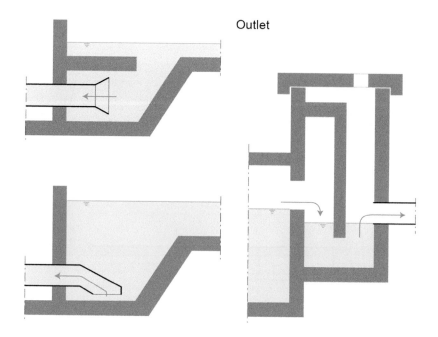

Figure 4.118 Reservoir outlet arrangements
Source: Obradović, 1992.

Regarding pumping stations, the main engineering aspects are (Williams and Culp, 1986):

- accessibility and layout of the station,
- foundations and vibration,
- drainage,
- acoustic insulation,
- heating, ventilation and lighting,
- corrosion,
- protection against vandalism, and
- health and safety.

The facilities for installation, servicing, dismantling and removal of pumps should be planned while keeping the size of the whole station at a minimum. The design of the pipe work should be flexible to accommodate tolerances in assembly, and also to maintain the units without stopping the operation of the whole station (Figure 4.119); possible future extensions should also be born in mind.

Well-constructed foundations are essential for good operation, particularly where the pump and motor are not integral. Where vibration is expected, foundation blocks should be isolated with damping material from the rest of the structure. Isolation of the piping from vibration should also be provided (Figure 4.120).

Most of the design requirements can be successfully tackled if the station is constructed partly or wholly in the ground. Savings in architectural finishes are likely to offset the additional costs of burying the structure, providing good hydraulic performance (NPSH) as well as sound, heat and frost insulation, protection

Figure 4.119 Piping in a pumping station

against vandalism, etc. However, the disadvantages are that access to the station is more difficult, the drainage system in the station needs additional pumping, condensation may be a hazard for electrical equipment, etc.

Developments in remote control and telemetry have virtually removed the need for stations to be permanently manned; the buildings for pumping stations are required primarily to accommodate and protect the machinery, providing its smooth operation and maintenance (Figure 4.121).

Figure 4.120 Horizontal pumps with rubber rings to dampen vibration

Figure 4.121 An unmanned booster pumping station

4.5.10   Standardisation and quality assessment

The need for standardisation is clear for water distribution companies:

- Regarding planning and design, it allows engineers to be more precise in drafting specifications. The required testing and inspection of materials guarantees the quality of applied materials.
- Regarding O&M, it reduces difficulties caused by the diversity of the installed material.

Hence, the first step towards standardisation is the elimination of variety within certain products. Thereafter, an evaluation of the technical characteristics of the component follows, namely the fitness for the purpose that must result in a sufficient and guaranteed quality level for the use of products.

The quality assessment, which is implicitly the result of standardisation, is very important in this phase. In any manufacturing process there will be a deviation within certain limits and the products have to be tested; variations in quality may not be acceptable.

The level and frequency of testing depends on several factors but predominantly on the degree of certainty with regard to the consequences caused by inadequate quality. There are basically three different principles (Van der Zwan and Blokland, 1989):

- batch testing,
- testing and assessment by the manufacturer's internal control, followed by surveillance, and
- internal assessment by the manufacturer.

By batch testing, a product sample is tested and a verdict on its compatibility with the specifications is issued.

The second principle is sample testing of the product according to a prescribed method approved by the manufacturer's quality control. This testing is regularly followed by surveillance in the form of quality control in the factory, but also by audit testing by both the factory and open market.

The third system concerns the manufacturer's capability to produce consistently in accordance with required specifications. The manufacturing methods, facilities and quality control are assessed and approved in respect of a technology that is capable of delivering products of constant quality.

It has to be realised that any quality control requires qualified personnel, and accurate measuring and testing equipment. The factory justifies the cost of such investment only if it increases productivity in the long term. On the other hand, the processes that can provide an initially low level of rejected products usually include modern technologies that are also expensive. Hence, the manufacturer is primarily interested in introducing as much quality control as necessary to achieve good sales figures. Moreover, the absence of real market competition or/and supervision may lead to the relations between manufacturers and water supply companies becoming based exclusively on mutual trust and willingness for cooperation. Two possibilities to prevent this are:

- the centralisation of testing, and
- a certification procedure.

Centralisation of testing can be obtained by establishing an institution that will assist the water sector with services related to quality control. This offers obvious advantages:

- lower expenses for quality control,
- availability of proper equipment at the required time,
- continuity of testing and inspection,
- development of experience in the field, and
- the exchange of information.

Certification of industrial performance is not only important for the water industry as a client, but it is also an assurance to the manufacturer that the required quality has been agreed and is acceptable for everyone. The institution responsible for centralised testing is also usually the one that issues certificates. It is the ultimate controller of the production and responsible for the quality in general.

Figure 4.122 KIWA-certified PVC pipes with the rated pressure of 0.75 MPa (75 mwc), D = 110 mm and wall thickness of 3.3 mm, produced in March 2009 from 'PVC with recycling warranty'

The experience of centralised control and certification has shown very positive results in the Netherlands, where these activities have been carried out by the KIWA institute since 1948. Nowadays, the vast majority of products for the water industry carry the KIWA label, which is practically a prerequisite for their implementation in any distribution network in the country. Examples of PVC pipes and household water meters are shown in figures 4.122 and 4.123, respectively.

Accredited by the Dutch Council for Certification, KIWA issues a large number of quality assurance certificates not only for waterworks products but also for building components, process and quality systems, etc. The certificates can be issued for:

- the quality of a product,
- an assessment of a manufacturing process,
- an assessment of an internal quality control scheme,
- an assessment of toxicological aspects of the production.

The certification procedure follows standard steps in each case, an example being given in Figure 4.124.

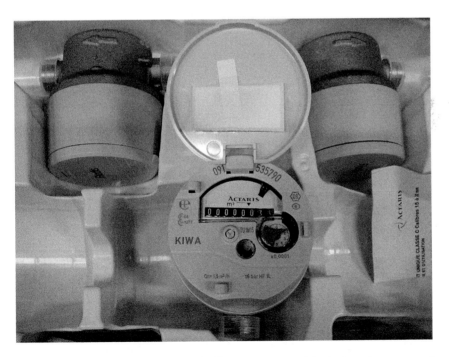

Figure 4.123 A KIWA-certified household water meter with the rated pressure of 16 bar (160 mwc), D = 15 mm and nominal flow Q_n = 1.5 m³/h, produced in 2009

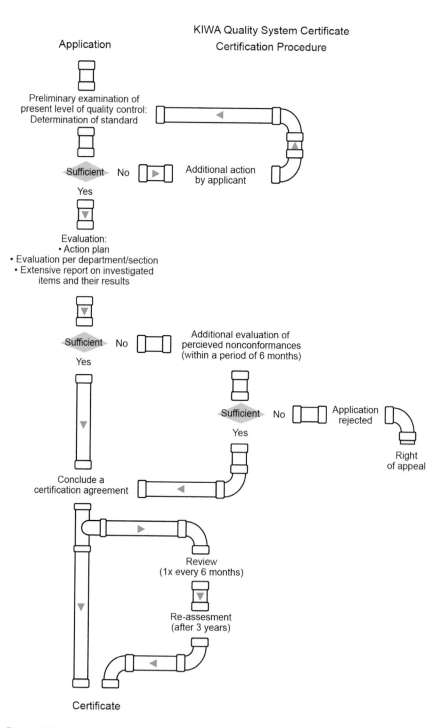

Figure 4.124 The certification procedure

Source: KIWA.

Chapter 5

Network construction

Network construction comprises the following steps:

1) site preparation,
2) excavation,
3) trench dewatering (if needed),
4) pipe laying,
5) jointing,
6) backfilling, and
7) testing and disinfection.

After the site has been prepared, all the other steps are conducted simultaneously at various sections of the pipe route; at its end, the pipes are tested; a few pipes further, the backfilling takes place; and at the same time at the preceding section the pipes are jointed, etc. This coordinated method of working is important in order to shorten the total duration of the construction, reducing both the cost and disturbance. The laying of a transportation steel pipe and a distribution PVC pipe are shown in Figure 5.1.

Figure 5.1 Pipe laying: a transportation steel pipe (left-hand side); a distribution PVC pipe (right-hand side)

Pipes can also be laid above ground or in tunnels, which then require adapted laying techniques such as the use of casings, anchorages and supports, etc. Some typical principles and solutions are briefly presented in this chapter.

5.1   Site preparation

Pipes can be laid only when the route is completely clear. Site preparation in urban areas can be a complex task where cooperation with other utilities is very important. Works on water, electricity, gas, road, or other infrastructure are often carried out simultaneously, as shown in Figure 5.2.

Before the work can commence, mutual agreement should be obtained about the working area so that other daily activities are not significantly affected during the construction. Proper signalling, footpaths and crossings for pedestrians, signs and warnings, a restricted access to the equipment in operation, etc. must be provided during the entire period of work. Water companies normally have responsibility and should be equipped to carry these tasks (Figure 5.3).

Pipes will be tested prior to leaving the factory and should also be tested after reaching the site in order to check for possible damage resulting from transportation. Further damage to the pipe is possible during the process of unloading, stacking and/or stringing along the laying route. The dropping of pipes, pipes striking each other, bundling pipes too high, or stacking them on an uneven surface or without proper support will all have a negative effect. A good example of stacking PE pipes is shown in Figure 5.4.

Each scratch on the external or internal coating of a metal pipe is a potential source of corrosion. Cement-based pipes are very vulnerable to impact damage

Figure 5.2 Simultaneous laying of drinking water, sewage, and gas pipelines

Figure 5.3 Signs for traffic regulation during network construction

Courtesy: Waternet, Amsterdam.

Figure 5.4 Proper stacking of PE pipes

Figure 5.5 Fenced DI pipe storage on a construction site

and plastic pipes, although lighter, are not an exception in this respect; scratches on PVC reduce the pipe strength. Hence, a final check is necessary for each pipe before it is put into position.

Pipes and fittings waiting to be installed should be kept clean in a fenced storage as a protection against potential theft and vandalism (Figure 5.5).

Before excavating paved surfaces and roads, the edges of the trench have to be cut to avoid damage to surrounding areas. If traffic loads allow, the pipe route will be located alongside the road, preferably not too far from it, which reduces damage to the pavement resulting from excavation. Breaking the surface is usually carried out by pneumatic hammers. Large pieces of concrete and asphalt will be removed from the site as they will not be used for backfilling. If the surface is not paved, the topsoil is usually removed by scrapers and stacked for use in the final reinstatement of the site.

5.1.1 *Excavation*

Excavation is the most expensive part of pipe laying. The choices of technique and trench dimensions are therefore very important factors that will affect the total cost. The preferred excavation method depends on:

- available space on the site,
- soil conditions, and
- width and depth of the trench.

Excavation is commonly carried out by mechanical excavators (Figure 5.6). In areas where there are obstructions (e.g. other services are in the trench) or access for the machinery is restricted (small streets, busy traffic, etc), excavation by hand might be required (Figure 5.7). For smaller trenches (up to 300 mm wide and 1 m deep) vacuum excavation can be used. Here, after breaking the surface and removing the top layer in the conventional manner, a special pneumatic digging tool is used. With this method, the soil is then removed through a flexible hose.

Care has to be taken during the work:

- to stabilise the walls, either by battering or shoring,
- to clear the trench edges of chunks of rock or earth that could potentially damage the pipe or injure the workers,
- to leave enough space between the trench and pile of excavated material, and
- to keep the work as dry as possible.

Figure 5.6 Mechanical excavation in sand: a wide trench (left-hand side) and a narrow trench (right-hand side)

Figure 5.7 Manually excavated trenches

Batter-sided trenches are rarely used in urban areas because of the space needed. Where possible, the angle of slope should depend on the trench depth and soil characteristics, as shown in Figure 5.8.

The choice of technique, dependant on the soil conditions, is often prescribed by laying regulations. Three groups of soils can be distinguished regarding their suitability for excavation (see Figure 5.9).

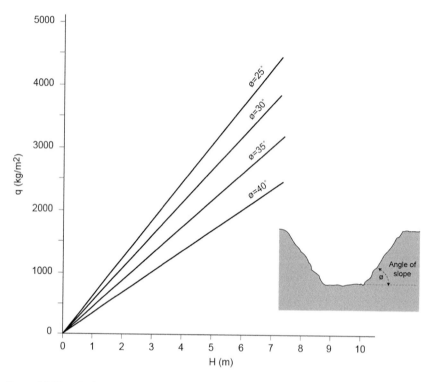

Figure 5.8 Trench slopes

Source: Pont-a-Mousson, 1992.

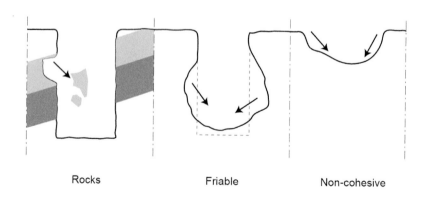

Rocks   Friable   Non-cohesive

Figure 5.9 Soil types

Rocks are extremely cohesive materials but the possibility of collapse cannot be excluded. Cracks are sometimes present, which can result in rocks falling. Excavation is difficult in this type of soil.

Friable soils are the most common soils. A certain degree of cohesion allows them to hold together for a while during excavation. However, these soils are very sensitive to water, and collapse of the trench walls caused by the vibration of the equipment is also possible.

Non-cohesive soils are soils without any cohesion (e.g. dry sand, mud or freshly restored backfill), which collapse almost instantly. Protection against the danger of collapse is therefore essential.

Regarding the soil type, AWWA (2016-1) suggests the slopes as shown in Figure 5.10.

To protect the trench from collapse, different techniques for *shoring* can be applied by:

a)   prefabricated wooden panels (jointed or single),
b)   wooden or metal sheets, and
c)   pile-driven sheets.

An example of shoring assembly by wooden panels is shown in Figure 5.11. Another example, done with pile-driven metal sheets, can be seen in Figure 5.12.

The *shielding* technique can be used in rocky and friable soils, in the absence of shoring. With this method, the laying and jointing work takes place in a partly open steel box (Figure 5.13) that is pulled throughout the trench as the work progresses. The sidewalls of the box do not prevent occasional caving in of the soil, as the width of the box is smaller that the trench width in order to be able to pull

Rocks

Friable soils

Non-cohesive soils

Shielding

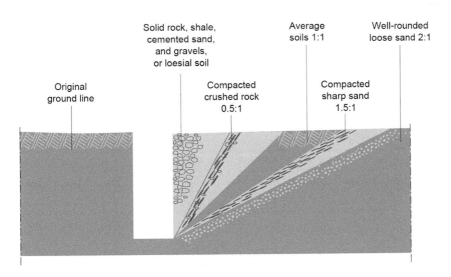

Figure 5.10 Required trench slopes for different soil types

Source: AWWA, 2016-1.

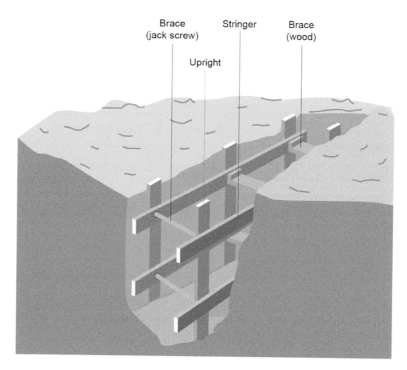

Figure 5.11 Shoring with wooden panels
Source: AWWA, 2016-1.

Figure 5.12 Shoring with pile-driven metal sheets

Figure 5.13 Steel boxes used for the shielding technique

it smoothly. The main objective here with this method is the protection of the workers.

How much trench is excavated depends on the time necessary for pipe laying and backfilling. Normally, the trenching is excavated a day or two ahead of the pipe laying, depending on the laying methods applied. However this should not be carried out too far in advance, as empty trenches may accumulate rainwater and are potentially dangerous, especially outside working hours.

The width of the trench at the bottom depends on the pipe diameter. An additional space of 0.3-0.6 m around the pipe (external diameter) should be provided for shoring and jointing works.

Extreme temperatures can have an impact on the operation of water distribution systems, not only by affecting the water consumption but also causing pipe damage either by freezing or very high temperatures. While deciding on the optimal trench depth, care should be taken to minimize the temperature impact on pipes and joints. On the other hand, increasing the depth beyond what is really essential is more costly, not only during installation but also in the maintenance phase. Some degree of pipe burst under extreme weather conditions is always acceptable if the repair can be conducted quickly and without disturbance to a large number of consumers. In general, the minimum cover over the pipe crown in moderate climates are:

*   1.0 m, for transmission lines,
*   0.8 m, for distribution pipes, and
*   0.6 m, for service pipes.

For frost prevention, pipes are laid deeper in areas with a cold climate, sometimes up to 2.5-3 m, which depends on the degree of frost penetration in the ground. An example of trenches as deep as 5 meters can be met in Mongolia (Figure 5.14, left-hand side). The reason is that this country suffers from extremely low winter temperatures running up to minus 50°C and resulting in the soil freezing down to a depth of over 4 meters. Alternatively, pipes in shallow trenches can be laid with thermal insulation; an example in Figure 5.14 (right-hand side) shows additionally insulated DI pipes. In extremely hot climates, the pipes will also be buried deeper, mainly to preserve the water temperature. Examples of typical depths of soil covers from European practice are shown in Table 5.1.

The excavated material is deposited alongside the trench if it is going to be used for backfilling. Its location should not be too far from the trench but also not too close, as this exerts pressure on the trench wall, risking its collapse. Moreover, it also limits the movement of the workers. In general, approximately 0.5 m space should be left free for the deposited material.

Tunnelling

Excavation for laying pipes passing under roads, railways and watercourses is done by *tunnelling*. The special reason for this is to protect the surrounding area from erosion caused by the pipe burst or leakage, which can have catastrophic

Figure 5.14 Trench depth (left-hand side) and insulated DI pipes (right-hand side) in Ulaanbaatar, Mongolia

Table 5.1 Soil cover over pipes

Country	Depth (m)
Austria	1.0–1.5
Belgium	0.8–1.0
Finland	2.1–2.5
Germany	1.1–1.8
the Netherlands	0.8–1.0
Switzerland	1.2–1.5

consequences. Secondly, the pipe is protected in this way from soil subsidence and vibrations caused by traffic, and maintenance can be carried out without interruptions or breaking of the surface.

Excavation of tunnels is a very expensive activity. In this situation *thrust boring* is applied, whereby a rotating auger moving the excavated material backward pushes a steel shield pipe forward. New lengths of pipes are welded or jointed together as the tunnelling proceeds, finally appearing at the other side of the crossing. The thrust boring technique is successful for short lengths of tunnels, up to 100 m, and for pipes of maximum 2500 mm diameter (Brandon, 1984).

Thrust boring

For longer lengths and larger diameters, a tunnel should be constructed by traditional methods. These structures can also serve to accommodate several pipes, usually water mains carrying large quantities of water. In rock, the tunnel section can be a vertical wall lined with concrete; for other soils circular sections formed by reinforced concrete segments are common. When the tunnel is shallow it can be constructed by the *cut-and-cover method* and in this situation a reinforced concrete box culvert is a more suitable solution.

Cut-and-cover method

An example of boring equipment for weaker soils is shown in Figure 5.15. With this technique, a thrust is applied to move the head (Figure 5.16, below) whose angle will be adjusted for precise (computer-controlled) direction and the depth of the pipe waiting on the downstream side to be pulled backwards. The required length is achieved by assembling additional sections of the drill (to be seen in Figure 5.16, above) until the downstream end has been reached (shown in Figure 5.17). In the application shown in figures 5.15-5.17, a steel pipe of 1000 mm

Figure 5.15 Boring in friable soils, upstream view

Figure 5.16 Drill sections (above) and the head (below)

Figure 5.17 Downstream view

Figure 5.18 Steel pipe, D = 1000 mm, ready for laying

Courtesy: Evides, Rotterdam.

diameter was laid under a river in 2007 (shown in Figure 5.18; courtesy: Evides, Rotterdam).

5.1.2 Trench dewatering

The normal method of removing water as it enters the excavation is by pumping (Figure 5.19). Sand and silt in unstable soils are mixed with water and carried out as well. However, if this continues over a period of time, there is a danger of subsidence in the adjacent ground. In such situations, the removal of groundwater can be carried out by using wellpoint dewatering equipment (Figure 5.20). The water is collected through perforated suction pipes put in the ground below the lowest excavation level. All suction pipes are connected to the header pipe, which transports the water by vacuum created by a wellpoint pump. Two applications of this method is shown in figures 5.21 and 5.22.

Although proven to be very efficient in the case of non-cohesive soils, the wellpoint dewatering method can rarely be used in impervious soils because the water is not able to flow to the extraction points. Instead, electro-osmosis, forcing the water by means of a passage of electrical current to a dewatering point, may be successful in maintaining vertical sides in wet unstable silt.

Figure 5.19 Trench dewatering by pumping

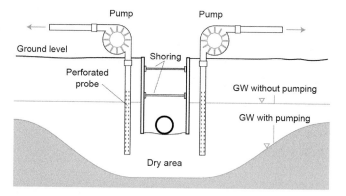

Figure 5.20 Principle of the wellpoint method

Figure 5.21 Application of the wellpoint method

Figure 5.22 A lateral pipe connected to the pump and the discharge box to the sewer

5.2   Pipe laying

5.2.1   Laying in trenches

The trench bottom provides the pipe's foundations. In homogeneous, even and well-consolidated soils, pipes can be laid directly on the bottom. The entire length of the pipe should touch the ground surface. To facilitate this, the space around the joints should also be excavated. In rocky soils, a pipe bed of 15-20 cm should be provided (Figure 5.23). Depending on the pipe material, the bed can be made of sand, gravel or dry concrete, which assumes that the surface of the trench bottom is even and properly compacted.

When it is necessary to lay on less stable ground, pipes should be supported on piles reaching a stable soil, if such soil is to be found at a depth less than a few meters. Care should be taken to avoid point loads being transmitted to the pipes (particularly in the case of PVC pipes).

Piles can also provide support to the pipes in waterlogged grounds. If this is not sufficient, lowering the groundwater table can be achieved by laying a drain alongside the trench at a depth of 0.5 m below the pipe invert. The pipe is then bedded on the reinforced concrete raft placed across the trench bottom, which ensures its stability.

An example of concrete transportation pipes laid on wooden piles is shown in Figure 5.24. However, laying the pipes on wooden piles carries risks because these may rot in the longer term. Their cracking may lead to a subsequent failure of neighbouring piles originating from the lack of sufficient support, ending in the burst of several pipes. This technique is therefore not advised in aggressive soils and in extremely cold temperatures.

Most pipes are still laid individually in the trench. With the increased use of flexible pipes, the technique of laying large sections of distribution mains is becoming more common. The placing of pipes on the prepared bed in a position

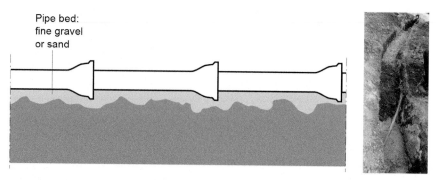

Pipe bed:
fine gravel
or sand

Figure 5.23 Pipe bed

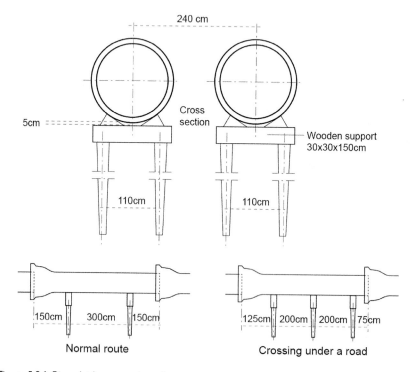

240 cm

Cross
section

5cm

Wooden support
30x30x150cm

110cm

110cm

150cm   300cm   150cm

Normal route

125cm   200cm   200cm   75cm

Crossing under a road

Figure 5.24 Pipes laid on wooden piles

ready for jointing requires appropriate equipment and skill (figures 5.25 and 5.26). The precise laying procedure depends on the pipe material; the advice of pipe manufacturers must be taken into account here. The entering of groundwater or rainwater into the pipeline is highly undesirable, so pipe stoppers should be used if the work has to be paused e.g. at the end of the day. In highly corrosive ground, metal pipes (and joints) can be sleeved into a polyethylene film as an additional protection to the external coating (Figure 5.27); an application for DI pipes is to be seen in Figure 5.28.

Figure 5.25 Testing of external coating

Figure 5.26 Pipe positioning in a trench

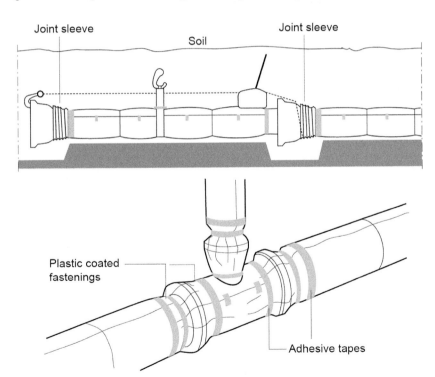

Figure 5.27 Protection of joints

Source: Pont-a-Mousson, 1992.

Figure 5.28 A sleeved DI pipe

5.2.2   Casings

The pipe shown in Figure 5.18 is laid by pulling it in the opposite direction to the boring. This laying technique is called *casing*. Different principles of casings are possible; the method typical for DI pipes is shown in figures 5.29 and 5.30.

   Old pipes can sometimes be used as casings for the new pipes (Figure 5.31). This solution will probably reduce the maximum conveying capacity, although

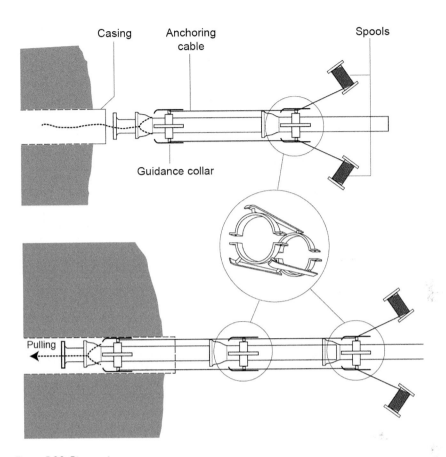

Figure 5.29 Pipe casing

Source: Pont-a-Mousson, 1992.

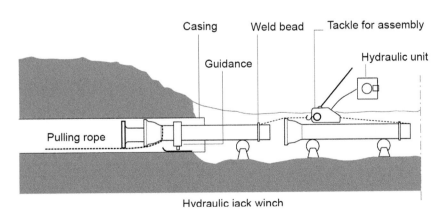

Figure 5.30 Pipe casing

Source: Pont-a-Mousson, 1992.

Figure 5.31 Casing of a PE pipe in an old CI pipe (left-hand side); the opposite view (right-hand side)

the smaller diameter is partly compensated for by the reduced roughness values of the new pipe. Special attention should be paid, in general for all casing choices, to the jointing of the new pipes in order to make the route leakage-free, because there will be no room whatsoever for any possible future repairs, once the pipe has been laid.

5.2.3   Laying above ground

The following aspects should be considered when laying pipes above ground:

a)   the design of the support system,
b)   the accommodation of thermal expansion,
c)   the anchorage of components subjected to hydraulic thrust, and
d)   protection against extreme temperatures (where necessary).

A few examples are shown in figures 5.32-5.35.

Figure 5.32 Steel pipes clamped on a concrete support and protected by a roof

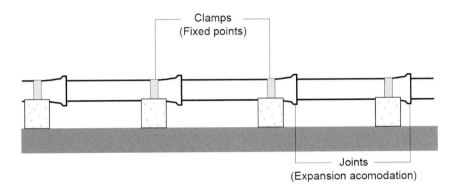

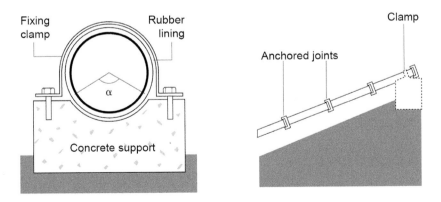

Figure 5.33 Pipe laying on a concrete support

Source: Pont-a-Mousson, 1992.

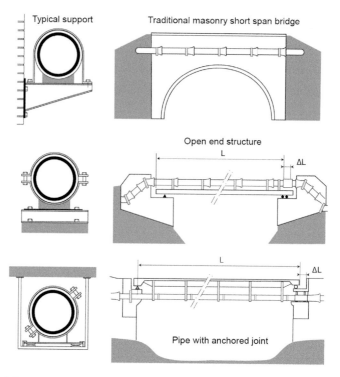

Figure 5.34 Pipe laying at a crossing

Source: Pont-a-Mousson, 1992.

Figure 5.35 Concrete pipe laying at a bridge

5.3 Pipe jointing

Examples of jointing principles and tools are shown in figures 5.36-5.40.

5.3.1 Flanged joints

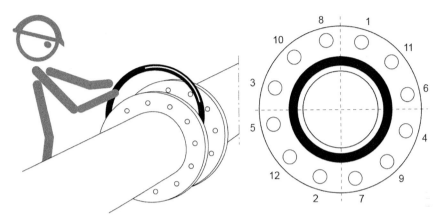

Figure 5.36 Pipe jointing using flanged joints

Source: Pont-a-Mousson, 1992.

5.3.2 Gland joints

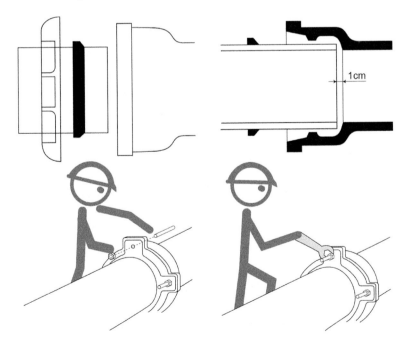

Figure 5.37 Pipe jointing using gland joints

5.3.3 'Push-in' joints

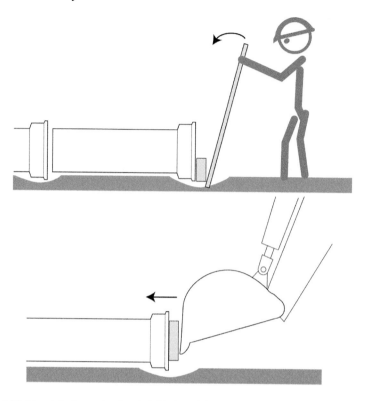

Figure 5.38 Pipe jointing using 'push-in' joints (above: manual push, below: mechanical push)

Figure 5.39 DI pipe jointing using mechanical push

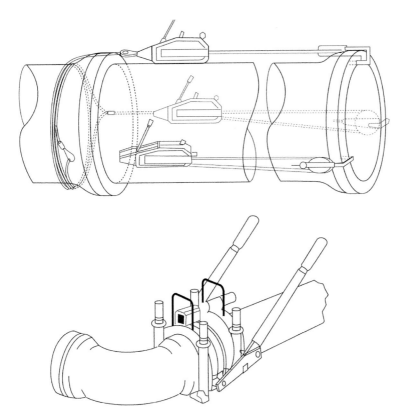

Figure 5.40 Jointing equipment for larger diameter DI pipes

Source: Pont-a-Mousson, 1992.

5.3.4 *Anchorages and supports*

After the pipes have been laid and connected, the concrete anchorage and support structures must be cast before backfilling is completed. Anchor blocks are designed depending on the pipe configuration and soil characteristics. The design takes into account the forces involved and the result is usually expressed as the volume of concrete required to carry the thrust (Figure 5.41). The water pressure taken into consideration for this calculation is the maximum anticipated pressure, with an additional safety factor in case there are pressure surges.

Concrete should be placed and consolidated against undisturbed soil and around the pipe or fitting to achieve a good bond. Care must be taken when filling with concrete to keep joints clean. The position of the thrust blocks for some typical bends and junctions is shown in Figure 5.42 (AWWA, 2003). In situations when the positioning of the blocks is difficult due to obstructions or the absence of undisturbed soil, a specific anchor, also applicable for vertical bends, can be used with the pipe connected to a concrete block with steel rods, as shown in Figure5.43 (AWWA, 2016-1).

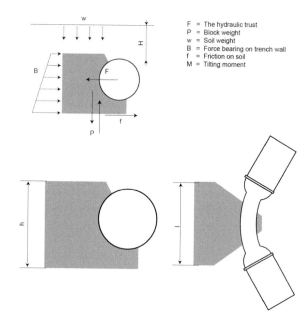

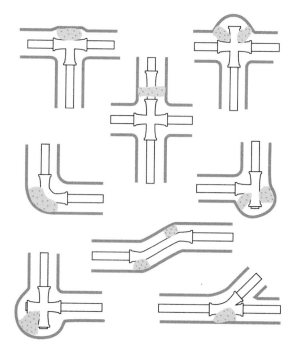

Figure 5.41 Anchorage of pipe bends

Source: Pont-a-Mousson, 1992.

Figure 5.42 Thrust blocks in distribution systems

Source: AWWA, 2003.

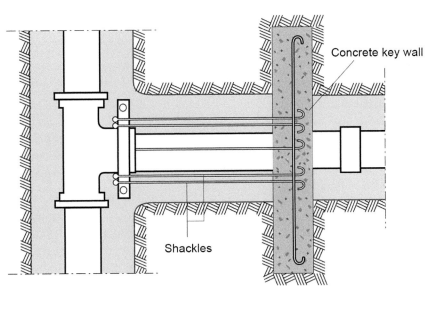

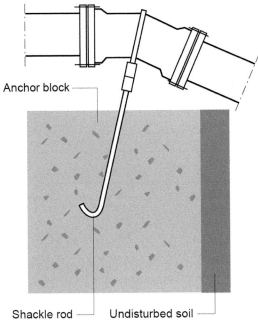

Figure 5.43 Anchorage of pipe bends

Source: AWWA, 2016-1.

5.3.5 Backfilling

Backfilling of the trench can be done in two phases: partly, immediately after the pipe laying to prevent floating caused by sudden heavy showers, and finally, after completion of the hydraulic tests. These two general layers shown in Figure 5.44 are:

1) pipe surround (initial backfill), and
2) main backfill (infill).

The surround provides stability and protection for the pipe and increases the bearing capacity for external loads. The type of material used depends on the pipe characteristics and soil conditions. The infill varies according to the area involved and stability of the surface.

Fine material should always be used for the initial backfill; excavated subsoil may also be suitable. Stones, rocks and any sharp materials are not allowed close to the pipe. The soil is normally placed in the trench in layers of 15-20 cm, and each layer is well compacted by machines that do not damage the pipe (Figure 5.45). The pipe can also be partly surrounded by the initial backfill but this reduces its supporting strength to a large extent; Table 5.2 illustrates this.

Top backfill in urban areas usually has to follow specifications required by the road authorities, and in open areas it is more related to aesthetics.

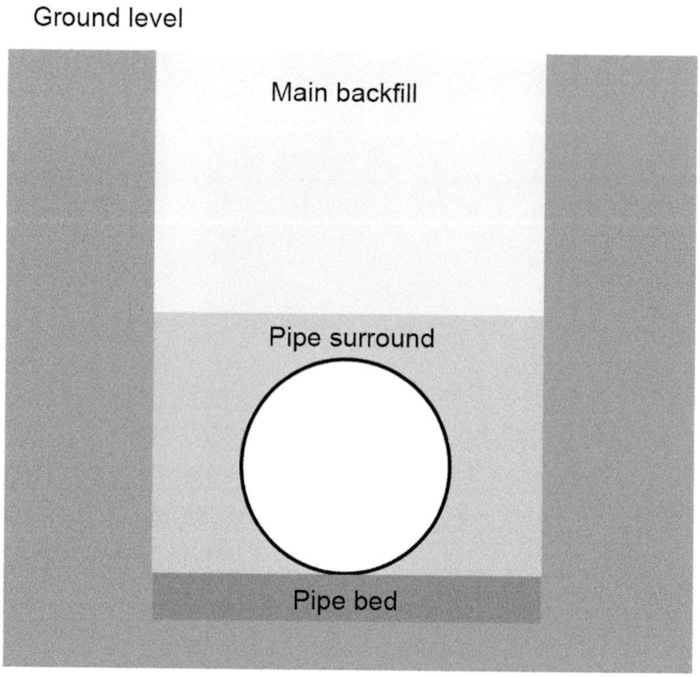

Figure 5.44 Backfill composition

Figure 5.45. Backfilling and compacting of the trench

Table 5.2 Load-bearing strength of rigid pipes

Degree of initial backfill	Increase in load-bearing strength (%)
No initial backfill.	-
Backfill up to 50% of horizontal diameter.	36
Backfill up to 60% of horizontal diameter.	73
Backfill up to full diameter (half pipe).	114
Backfill covering entire pipe (Figure 5.44).	150

Source: AWWA, 2003.

5.3.6 Testing and disinfection

As soon as the pipe laying is completed, a hydraulic test has to be carried out to check the quality of workmanship, namely:

- the mechanical strength and leak tightness of the system, and
- the strength of the anchorage and support structures.

All changes of directions, fittings and valves should be permanently anchored before the test starts. The ends of the tested section must be securely closed and temporarily anchored as well. There must be sufficient backfilling to prevent movement of the pipes during the test, but the joints should be left exposed until testing has been completed (Figure 5.46).

Water mains can be tested in lengths varying from a few hundred metres up to about a kilometre. However, although any length is possible in theory, in practice it is more difficult to detect leaks with distances of more than 500 m. It is sometimes possible or desirable to carry out a double test, first on longer and later on shorter sections. The test pressure applied depends on the regulations. For distribution systems, it is usually 50 % higher than the maximum working pressure expected in the network. The test pressure also needs to include a compensation of the static head generated by the elevation difference between the beginning and the end of the test section, and will be even higher if surge pressures are expected during the network operation. A common testing installation is shown in Figure 5.47.

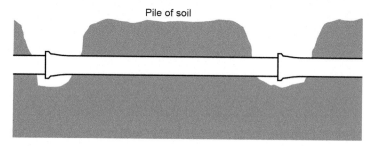

Figure 5.46 Preparation for pipe testing

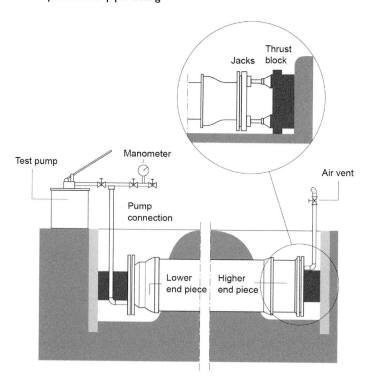

Figure 5.47 Pipe-testing equipment

Source: Pont-a-Mousson, 1992.

The test starts by filling the section with chlorinated water, if possible from the lower of the two pipe ends. It is essential to ensure that the main has been completely purged of air before it is pressurised, and also that the valves on possible pipe branches are closed. Removing the air completely is however not always easy and some tolerance is acceptable with this respect. The standards in UK prescribe the upper limit of 4 % (UK Water Industry, 2015).

After filling, the section should be left under moderate pressure until stable conditions are achieved. The length of this period depends on the quantity of the air trapped and the absorption of the pipe material. For absorbent pipes such as AC and concrete, or cement-lined pipes, it can take a couple of days before the pipe material is fully saturated. Plastic pipes (PVC and PE) can also lose the pressure during the similar period due to creeping of the material (weakening of the molecular structure as a result of significantly increased pressure). These specific conditions may lead to modifications of the testing procedure for different pipe materials. The saturation time in reality will be kept much shorter, from a few dozen minutes to a few hours, with corresponding corrections in the formulas applied to calculate expected results to pass the test.

The pressure is then brought up to the test value by a pump and all the exposed parts of the section are examined for water tightness. The duration of the test and interpretation of the results depend on the regulations. According to French standards for DI pipes, the test is successful if the pressure in the section does not drop more than 2 mwc within 30 minutes (Pont-a-Mousson, 1992). In British standards, a leakage level in the section is monitored through the amount of water pumped to re-establish the testing pressure after the drop. Tolerable leakage levels per pipe diameter are shown in Table 5.3 (UK Water Industry, 2015).

Table 5.3 Allowable leak rates as a function of diameter

Nominal pipe diameter (mm)	Leakage rate (l/km/h)
100	0.18
150	0.41
200	0.72
250	1.13
300	1.62
350	2.21
400	2.88
450	3.65
500	4.50
600	6.48
700	8.82
800	11.52
900	14.58
1000	18.00

Source: UK Water Industry, 2015.

If these limits are exceeded, a systematic search for leaks must be made. If standard methods of leak detection do not produce a result, the testing has to be repeated on shorter sections in order to isolate the leakage points. On rare occasions air pressure testing can be used for locating defective joints in waterlogged conditions.

When the hydraulic test has been successfully completed, the pipeline should be flushed out to remove any remaining debris and properly disinfected, if necessary. Before being washed out, the water in the pipeline must be de-chlorinated and then carefully disposed of to a pre-planned site. After washing out, the network can be charged with water, and after testing the water quality, it can be put into service.

Pipe testing requires trained personnel and strict adherence to all the safety measures. Working with high pressures requires a properly isolated site in order to minimise the consequences of possible calamities. Furthermore, depending on the situation, pipe testing can be a time-consuming process. Therefore, because the quality of jointing has significantly improved nowadays, it is not unusual for water supply companies to omit this step and complete the pipe laying by only flushing and disinfecting.

Chapter 6

Network operation and maintenance

6.1 Network operation

The consumers' requirements will not be satisfied in a network that is poorly operated, even if it has been well designed and constructed. Making errors in this phase amplifies the common problems and their implications that have already been mentioned in previous chapters:

- low operating pressures causing an inadequate level of service,
- high operating pressures causing high leakage in the system,
- low velocities causing long retention of water in pipes and reservoirs,
- high velocities unnecessarily increasing the pressure losses (i.e. the energy consumption needed for conveyance), as well as the risk of surge pressures, and
- frequent changes of flow direction causing water turbidity.

Water networks can be compared with living organisms where several complex processes interact, as can be seen in Figure 6.1 (Vreeburg, 2007).

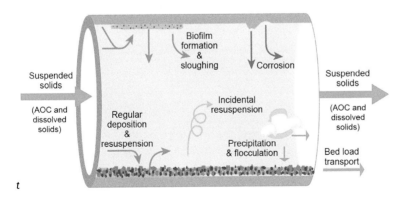

Figure 6.1 Particle-related processes in drinking water pipes

Source: Vreeburg, 2007..

Drinking water is never perfectly pure and can convey traces of assimilatable organic carbon (AOC) and various suspended/dissolved solids originating from an inadequate treatment process and/or intrusion of pollution through damaged pipes and joints. Specific hydraulic operation may enhance the deposits of sediments, their resuspension, formation of biofilms, and corrosion growth or release, all of which cause discolouration of water, bad taste and odour, and in extreme cases have a serious impact on public health. Coping with these problems on a regular basis influences maintenance requirements and the overall exploitation costs. Alternatively, managing the operating pressures and velocities is often a cheaper measure and therefore one of the first taken to mitigate the water quality and water loss problems in transport and distribution networks.

The hydraulic operation of gravity systems is relatively simple and deals with the balance between demand and supply, which can be controlled by operating valves. Pressure limitations in gravity systems that result from topographic conditions become even bigger in the case of a bad design. A wrongly elevated tank, incorrect volume or badly sized pipe diameter will not guarantee optimal supply, and errors will have to be corrected by what would otherwise be unnecessary pumping.

In pumped systems, a more sophisticated operation has to be introduced to meet the demand variations and keep the pressures within an acceptable range. Computer simulations are an essential support in solving problems such as these. As well as pressures and flows, network models can process additional results relevant to the optimisation of the operation, such as: power consumption in pumping stations, demand deficit in the system, the necessary storage volume, or decay/growth of constituents in the network. These are also able to describe the patterns developed during irregular supply situations. Finally, the models can be linked to monitoring devices in the system, which enables the whole operation to be coordinated from one central place.

To remain effective, computer models need to be continuously calibrated using telemetry results. An example in Figure 6.2 shows a close compliance between the computer simulation results and telemetry measurements ('+' markers) for the demand and the pressures in an urban district with a population of nearly 6000 (1270 connections) located north of Colombo in Sri Lanka, and supplied by gravity (Awyaja, 2015).

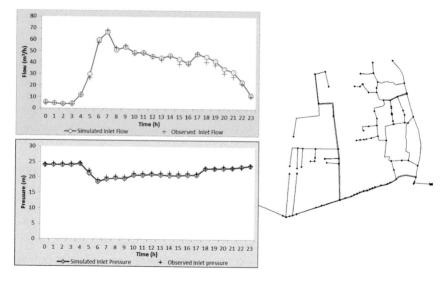

Figure 6.2 Fitting of telemetry and computer model results

Source: Awyaja, 2015.

In Figure 6.3, an example of a pumping regime controlled automatically by the water level variation in a corresponding tank is presented in part of the distribution network of Ulaanbaatar in Mongolia (Orgilt, 2008). This example clearly indicates the hydraulic link between a pumping station (link 290) and the neighbouring tank (node 34) that is filled with pumped water until its maximum water depth of 3.5 m has been reached. This begins at 5:00 hours when the tank starts supplying the area, and continues until 17:00 hours when the pump is again automatically switched on after the tank level drops to the specified minimum of 1.0 m, which indicates the emergency reserve.

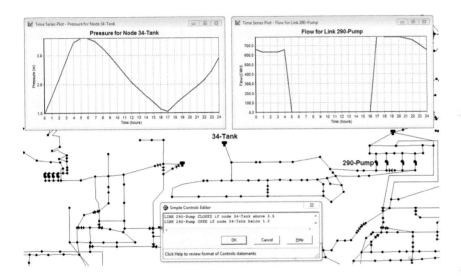

Figure 6.3 Simulated operation of a pumping station and corresponding tank

Source: Orgilt, 2008.

A computerised operation does not necessarily require lots of expensive equipment when compared to the overall investment cost of the distribution network. Maintaining this equipment in good condition is more of a concern, especially if it operates under extreme temperatures, humidity, unreliable power supply, etc. Nevertheless, thorough knowledge of the hydraulic behaviour of the system combined with well-organised work by sufficiently trained personnel can surely help to save on both the investment and the running costs.

6.1.1 *Monitoring*

Monitoring of water distribution systems provides vital information while setting up their operational regimes. It predominantly comprises:

- monitoring of variations in pressure, water level and flow (demand),
- monitoring of water quality parameters, such as temperature, pH, turbidity and colour, conductivity, chlorine residuals, etc.

Variations in pressure, level and flow/demand can be observed periodically for specific analysis (e.g. leakage surveys, the determination of a consumption pattern, or for billing purposes). When monitored continuously, they may indicate:

- operational problems that require urgent action (e.g. pressure drop due to a pipe burst),
- need for change in the mode of operation, as is the case shown in Figure 6.3.

Apart from the main concern of public health, the monitoring of water quality parameters can also help to detect inappropriate operational regimes. In addition, water quality parameters outside the normal range often indicate a need for necessary maintenance (an example is shown in figures 6.61 and 6.62 illustrated later in this chapter). As with hydraulic measurements, the selection of sampling points should provide a complete overview of the whole system; these will preferably be at the source, reservoirs, pumping stations and other easily accessible locations in the network where long retention times are expected.

Final decisions on the spatial distribution of measuring points depend on the configuration of the system. Meters for pressure and flow have to be installed in all the supply points and booster stations. Water levels in the reservoirs should also be permanently recorded. The measurements in main pipelines may be registered at critical points in the system (relevant junctions, extreme altitudes, pressure-reducing valves, system ends, etc.). All these data can be captured in one of the following ways:

Telemetry

1) *Telemetry* - There is a permanent online communication between the measuring device and control command centre where the parameter can be monitored round the clock.

Data loggers

2) *Data loggers* – Here a measuring device is permanently installed but the data for certain time intervals are captured periodically and will be processed and analysed later. Hence, the results of the measurements are not directly visible.

Local reading

3) *Local reading* - Direct readings can be obtained from the measuring devices' display and immediate action taken if required.

Sampling

4) *Sampling* - The water sample will occasionally be taken and transported to a laboratory, giving the option of extended analysis; these commonly include the determination of disinfectant residuals, coliform count, and heterotrophic plate count (HPC) as an indicator to determine the microbiological condition of water distribution systems i.e. as an early warning system for potential problems.

An example of a permanent monitoring station in the city of Amsterdam installed on a transmission main is shown in Figure 6.4. These real-time measurements can be seen in the control room for the entire network, which collects all the important information on supply, demand, pressures, levels of storage, status of pumping stations and important water quality parameters, stored in compact media that can be easily inspected even on an ordinary laptop. Figure 6.5 shows the control screens at the command centre of the Amsterdam water supply network; the system is practically fully automated and the visual control of the monitored parameters is merely for emergency purposes.

Figure 6.4 An on-line monitoring station at a fixed location

Courtesy: Waternet, Amsterdam.

Figure 6.5 Control room for Amsterdam's water supply network

Courtesy: Waternet, Amsterdam.

The database also contains historical data that can be retrieved on request, as the example of weekly supply patterns in the production facilities shown in Figure 6.6 illustrates.

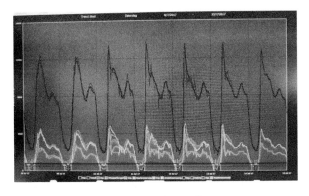

Figure 6.6 Weekly supply patterns in July 2017 in Amsterdam

Courtesy: Waternet, Amsterdam.

Battery-operated mobile containers equipped with a wider scope of monitoring equipment can occasionally be installed for a few weeks/months for temporary measurements (Figure 6.7). Lastly, mobile units can be connected to hydrants when there is a specific need for quick measurements of the main water parameters (Figure 6.8).

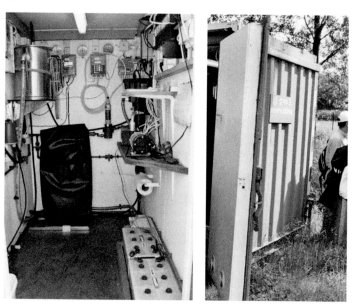

Figure 6.7 A mobile container for temporary water measurements

Figure 6.8 A portable kit for quick water measurements

Table 6.1 gives the recommended order of priority in selecting an appropriate data-capturing method in a water supply system (Obradović and Lonsdale, 1988).

Table 6.1 Data-capturing methods and points

Monitoring point	Flow	Water level	Pressure	Pump status	Pump speed	Valve opening	Volume	Chlorine	Turbidity
Water source	T/H	T/M							T/M
Well pumps	T/M			L/M	L/M				
Treatment plant	T/H	T/H					T/M		T/M
Main reservoir	T/M	T/H					L/H		
Main PST	T/H		L/H	T/L	T/M		L/H		
Local PST	T/H		L/H	T/L	L/M		L/M		
Service reservoir	T/M		L/H				L/M		
Booster PST	T/M			T/L	L/M		L/M		
Control valve	T/L		T/L			T/M			
Shut-off valve	L/H					L/H			
Distribution area	T/L		T/L				D/H		
Supply zone	D/H						D/M		
Demand district	D/H						D/L		
Control node	D/M		T/H					T/M	
Special customer	T/H						D/H	S/	
Large customer	D/H						D/M	S/	
Ordinary customer	D/M						D/L	S/	

Source: Obradović and Lonsdale, 1998.

X/Y: Method/priority
X: T - Telemetry D - Data loggers L - Local instruments S - Sampling
Y: H - High M - Medium L - Low

Special maintenance conditions in a system may offer opportunities to draw conclusions about its operation. The following example from the Netherlands (Cohen *et al.*, 1994) shows the monitoring of the retention times by measuring natrium concentrations at different points in the network. Retention times of up to 60 hours were observed in a distribution area near Amsterdam during maintenance of its main softening installation (Figure 6.9). The installation was stopped for 48 hours, which caused a temporary drop in concentration. Obviously, this effect could be registered sooner in the points located closer to the source (figures 6.10 and 6.11).

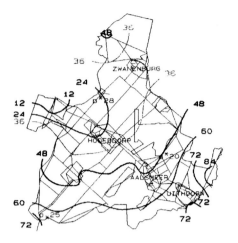

Figure 6.9 Retention times in a distribution network (hours)
Source: Cohen *et al.*, 1994.

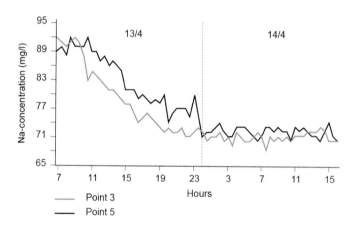

Figure 6.10 Drop in Na concentration as a result of hard water in the system
Source: Cohen *et al.*, 1994.

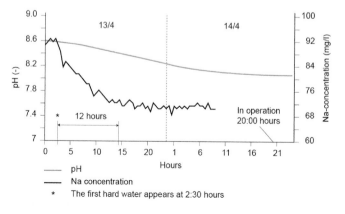

Figure 6.11 Comparison between Na concentration and pH values in the system
Source: Cohen *et al.*, 1994.

6.1.2 Network reliability

A network is reliable if it can permanently perform in accordance with its design criteria. In reality, due to unforeseen events, this is never the case. It is therefore more realistic to define network reliability as *the probability of a guaranteed minimum level of service in any (also irregular) situation.*

Water distribution literature contains a multitude of reliability definitions, as well as widely but not always consistently used terminology that describes distribution networks as *resilient, robust* or *redundant*. To clarify these, Lansey (2012) cites the general definitions of the US National Science Foundation as follows: '*Infrastructure resilience is the ability to gracefully degrade and subsequently recover from a potentially catastrophic disturbance that is internal or external in origin.*', while '*The robustness of a system to a given class of disturbances is defined as the ability to maintain its function when it is subject to a set of disturbances of this class.*' The set of disturbances that affect water system resilience typically comprises mechanical failures, power outages, human-induced failures, or natural disasters (earthquakes, for instance).

Resilience

Robustness

Various safety provisions, e.g. parallel/looped pipes, are *redundancies* that make water supply systems robust, both contributing to the system *availability*. Yet, it is not always that more redundancy results in improved robustness; inadequately planned provisions, e.g. a network layout that has added loops but the pipes are of too small a diameter, can actually make the system less robust.

Redundancy
Availability

Secondly, a more robust system is not necessarily more resilient. A distribution network may be highly resistant to a particular failure but may not recover quickly from it in the unlikely event that it does actually fail. This is for instance the case when a pipe burst takes too long to detect and repair because of poor monitoring or a poorly trained and equipped emergency response team. These aspects involve an additional two terms according to Lansey (2012): *resourcefulness* and *rapidity*. Hence, the network resilience rests on the so-called *4Rs* (redundancy, robustness, resourcefulness and rapidity) which all need appropriate planning in order to ensure reliable service levels.

Resourcefulness
Rapidity

When interruptions occur, consumers are normally not concerned with the cause, but rather with the consequences. Accordingly, the irregular events can be classified as *failures, calamities* or *disasters*. In the case of failures, this will cause a local interruption in the supply area. These failures are usually breaks in the distribution pipes that can be repaired within 24 hours. Failure of some major system component (a pumping station, or main transmission line) is considered to be a calamity, which will affect a larger number of consumers and in most cases for more than 24 hours. An event involving the simultaneous failure of major components is treated as a disaster. Temporary shortages of supply usually appear due to:

Failures
Calamities

Disasters

• pipe breakage caused by various factors,
• power or mechanical failure in the pumping station,
• detected water quality problems in the network,
• excessive demand caused by irregular/emergency events,
• maintenance or reconstruction of the system, and
• deterioration of the raw water quality at the source.

Modelling of pipe bursts

Pipe breakages are difficult to prevent because of the range of potential causes, shown in Figure 6.12, which impacts the (poor) pipe condition.

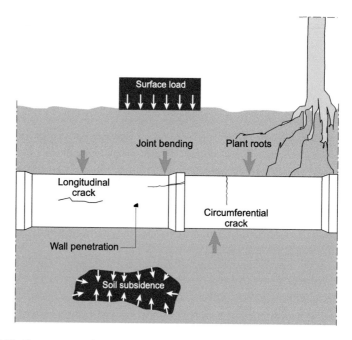

Figure 6.12 Common pipe burst types and causes

Table 6.2 shows pipe burst statistics for a number of European cities (main pipes only). The historical data in it paint a different picture depending on their presentation. For instance, three times the number of bursts are observed in the network of Vienna compared to Zurich, while at the same time the number of leaks per 100 km of the network is higher in Zurich than in Vienna. Thus, the network size obviously plays as important a role as the ability of the water company to carry out the repairs quickly and with minimal interruption in the service level.

Table 6.2 Pipe burst occurrences

City – country	Total bursts per annum	Bursts per 100 km pipes
Vienna - Austria	1700	47
Antwerp - Belgium	225	13
Copenhagen - Denmark	100	12
Helsinki - Finland	116	14
Frankfurt - Germany	600	40
Rotterdam - The Netherlands	400	19
Oslo - Norway	320	27
Barcelona - Spain	2850	115
Zurich – Switzerland	550	56

Adapted from: Coe, 1978.

The bursts occur more often with smaller pipes and service connections, but have a relatively insignificant impact on the overall water loss and hydraulic behaviour of the system. Reasonably accurate predictions can be derived from local observations of the event occurrence. The relation between the number of bursts and pipe diameters will normally follow similar trends to those shown in Figure 6.13.

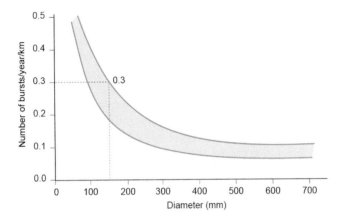

Figure 6.13 Bursting frequency of pipes of various diameters

With regard to the type of materials, in general pipes that have been attacked by corrosion are the most vulnerable. However, these experiences are not transferable in practice, and keeping local records about the system failure is an essential element of reliability analysis. The records for the three types of widely used pipe materials in the Netherlands in 1994 are given in Figure 6.14.

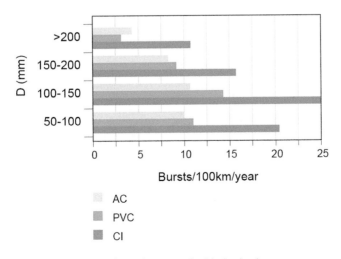

Figure 6.14 Average frequency of pipe bursts in the Netherlands

Source: Vreeburg et al., 1994.

According to the above diagram, CI pipes appear to be the most critical; in this particular case for well-known reasons. These are the first generation pipes, currently sometimes even 100 years old, while AC and PVC pipes belong to the second and third generations of the twentieth century, up to roughly 60 and 40 years old, respectively.

Given the total network lengths and variety of conditions (different material, size, age, type and maintenance practice), the asset management objective requires pipes to be analysed according to mechanical failure in order to predict breakage rates in the future. According to Watson et al. (2001), the modelling of pipe failures can be mainly grouped into three categories: (1) survival analysis, (2) probabilistic predictive models and (3) aggregated (regression) models. The first two categories deal with financial analysis to determine the most effective asset management strategy where the methods rely on a proper estimate of the economic lifetime i.e. the optimal timing for replacement. These approaches are highly dependent on the individual pipe characteristics due to the lack of long-term failure records.

Aggregated (regression) models

In the *aggregated (regression) models*, pipes with the same intrinsic properties are grouped together and linear regression is then used to establish a relationship between the age of the pipe and the number of failures. Shamir and Howard (1979) proposed an exponential model for pipe failure rate increase over time (also quoted by Engelhardt et al., 2000):

$$\lambda(t) = \lambda(t_o)e^{A(t-t_o)} \tag{6.1}$$

where, $\lambda(t)$ is the average annual number of failures per unit length of the pipe surveyed at year t, t_o is the base year for analysis, and A is the growth rate coefficient between years t_o and t. This approach does not provide any information about the variability that may exist between individual pipes in general. Consequently, Su et al. (1987) proposed a regression equation that correlates the failure rate λ and pipe diameter D using the data from the 1985 St. Louis Main Break Report (also quoted by Gargano and Pianese, 2000):

$$\lambda = \frac{0.6858}{D^{3.26}} + \frac{2.7158}{D^{1.3131}} + \frac{2.7685}{D^{3.5792}} + 0.042 \tag{6.2}$$

where D is the pipe diameter in inches and λ is the failure rate in breaks/mile/year.

Lifetime distribution models

Furthermore, the probability of pipe failures can be analysed by *lifetime distribution models*. For a well-designed and well-installed water distribution network, the series of pipes can be considered repairable because the cost of failure is small in comparison to their replacement. The main reasons for this are that firstly, pipes are mostly constructed under the ground and possible environmental effects of a failure are minimal. Secondly, storage tanks available within the network will keep a buffer volume for emergency situations. And thirdly, the networks commonly contain redundancy planned in their layout.

Repairable components typically have a 'bathtub'-shaped intensity function. This is shown in Figure 6.15 (adapted from Neubeck, 2004), where $\lambda(t)$ indicates the failure rate.

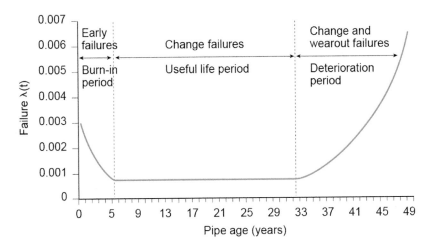

Figure 6.15 Pipe failure rate as a function of age

Adapted from: Neubeck, 2004.

Immediately after a pipe has been put into operation, the failure rate can be high due to poor transportation, stacking or workmanship during the installation. After early faults have been corrected, the intensity of bursts will decrease and remain relatively constant for long periods of the pipe's useful life. Following this period, the pipe will start to deteriorate faster, and the intensity of bursts will increase again. These bursts are considered as wear-out failures.

Two approaches are suggested in the literature to model the failure lifetime distribution: (1) Homogeneous Poisson Process (HPP) and (2) Non-Homogeneous Poisson Process (NHPP). The HPP neglects the time component of the failure and as such is mostly appropriate for renewable systems where repairs are executed regularly and the age of the pipe is within the useful life period. According to Shinistane et al. (2002), the probability of a pipe failure using the *Poisson probability distribution* is:

(Non) Homogeneous Poisson Process

$$p_j = 1 - e^{-\beta_j}, \text{ where } \beta_j = \lambda_j L_j \qquad (6.3)$$

Exponent β_j is the expected number of failures per year for pipe j, and λ_j and L_j its failure rate and length, respectively. Goulter and Coals (1986) and Su et al. (1987) also used similar methods with the HPP model to determine the probability of failure of individual pipes.

On the other hand, the NHPP model considers the time component and is therefore suitable for the burn-in and deterioration periods during which the intervals between the failures are neither independent nor identically distributed. The NHPP model also assumes negligible repair times, meaning that the repair time will have no effect on the increase in the pipe failure rate. This is considered acceptable when comparing a pipe lifetime measured in years with repair times counted in hours (Watson et al., 2001).

Tobias and Trindade (1995) describe the NHPP model in two versions:

(1) The power relation model: $\lambda(t) = \dfrac{dM(t)}{dt} = abt^{b-1}$ $\qquad$ (6.4)

(2) The exponential model: $\lambda(t) = e^{c+bt}$ $\qquad$ (6.5)

In the above equations, $\lambda(t)$ is the pipe failure rate at time t, $dM(t)$ is the expected number of failures between times 0 and t, and a, b, and c are empirically determined parameters from the historical burst records.

Although the probabilistic aspect of pipe failures have been incorporated in many reliability studies, the underlying assumption is that successive pipe failures can be modelled using HPP with the constant failure rate, λ. Consequently, Equation 6.3 has been widely used in reliability modelling of repairable systems.

Reliability assessment methods

Ostfeld (2004) classifies the approaches to the reliability assessment of water distribution systems into three groups: (1) analytical (connectivity/topological), (2) heuristic (entropy), and (3) simulation (hydraulic) approaches.

Analytical approaches deal with the layout of the water distribution network, which is associated with the probability that a given network keeps physically connected, given its component reliabilities. These are the approaches linked to the concepts of connectivity and reachability, not based on hydraulic simulations.

Heuristic approaches reflect the redundancy using entropy measures, as a surrogate for network reliability. Here, the key question is what a given level of entropy means in terms of reliability for a given distribution system.

Simulation approaches analyse the hydraulic performance of the network, i.e. the conveyance of the desired quantities and qualities of water at the required pressure to the appropriate locations at any given time. These approaches rely heavily on hydraulic models and require accurate information about the network layout and operation, including records related to the component failures. Owing to the availability of powerful computers and software, nowadays these are the most frequently explored approaches.

Simulation approaches

The pioneering hydraulic simulation approaches for network reliability analysis were based on demand-driven (DD) calculations, taking into consideration the pressure drop resulting from the failures. A practical method of this type is suggested by Cullinane (1989), who defined the nodal reliability as a percentage of time in which the pressure at the node is above the defined threshold. It reads as follows:

$$R_j = \sum_{i=1}^{k} \frac{r_{ij} t_i}{T}$$ $\qquad$ (6.6)

where R_j is the hydraulic reliability of node j, r_{ij} is the hydraulic reliability of node j during time step i, t_i is the duration of time step i, k is the total number of the time steps, and T is the length of the simulation period. Factor r_{ij} takes value 1 for the nodal pressure p_{ij} equal or above the threshold pressure p_{min}, and $r_{ij} = 0$ every time $p_{ij} < p_{min}$. For equal time intervals, $t_i = T/k$.

The reliability of the entire system consisting of n nodes can then be defined as the average of all the individual nodal reliabilities:

$$R = \sum_{j=1}^{n} \frac{R_j}{n} \qquad (6.7)$$

The above equations assume that all the network components are fully functional, which is rarely the case. The expected value of the nodal reliability can therefore be determined as:

$$RE_{jm} = A_m R_{jm} + U_m R_j \qquad (6.8)$$

where RE_{jm} is the expected value of the nodal reliability while considering pipe m, A_m is the availability of pipe m, i.e. the probability that this pipe is operational, U_m is the unavailability of pipe m, $U_m = 1 - A_m$, R_{jm} is the reliability of node j if link m is available, i.e. in operation, and R_j is the reliability of node j if link m is not available i.e. not in operation. The component availability can be calculated on an annual basis from the following equation:

$$A_m = \frac{8760 - CMT - PMT}{8760} \qquad (6.9)$$

where CMT represents the annual corrective maintenance time in hours and PMT is the annual preventive maintenance time in hours. These figures should be available from the water company records.

The values of R_{jm} and R_j in Equation 6.8 are determined from Equation 6.6, running the network computer simulation once with the link m operational and then again, by excluding it from the layout. The final outcome of the reliability assessment will be a coefficient that, for a given level of demand, takes a value between 0 and 1. As a consequence, the availability index of a distribution system will always be higher than its reliability index, as Figure 6.16 shows (adapted from Tung, 1996). Parallel to the mechanical availability, some researchers express the hydraulic reliability as *the hydraulic availability*, pointing to the availability of (sufficient) demands and pressures in the system.

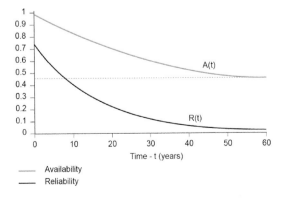

Figure 6.16 Reliability and availability trends
Adapted from: Tung, 1996.

No matter how practical the pressure-drop approach was, demand-driven simulations are not able to assess efficiently the loss of demand during failures which follows the loss of pressure. More meaningful results can therefore only be reached after the introduction of computationally powerful pressure-driven demand (PDD) simulations.

PDD simulations are also used in network reliability assessments to analyse the demand reduction inflicted by a single failure of subsequent pipes. The overall network reliability factor can be determined in general by a simplified formula:

$$R = 1 - \frac{Q - Q_f}{Q} \tag{6.10}$$

where Q_f represents the available demand in the system after the pipe failure, against the original demand Q. A more complex approach suggested by Ozger and Mays (2003) takes into consideration the probabilities of the pipe failures; the system reliability is expressed in terms of *available demand fraction* (ADF) under the minimum required service pressure:

$$R = 1 - \frac{1}{m} \sum_{j=1}^{m} (1 - ADF_{net}^j) P_j \tag{6.11}$$

where ADF_{net} is the network available demand fraction (i.e. the percentage of regular demand that can still be supplied) resulting from the failure of pipe j, P_j is the probability of the failure of that pipe, and m is the total number of pipes in the network. In the above equation, P_j is determined by using the Poission probability distribution described by Equation 6.3. In a calculation of the system availability, this method considers first and second order failures:

$$A = ADF_{net}^0 MA + \sum_{j=1}^{m} ADF_{net}^j u_j + \sum_{j=1}^{m-1} \sum_{k=j+1}^{m} ADF_{net}^{jk} u_{jk} \tag{6.12}$$

where ADF^0 is the available demand fraction with a fully functional network, ADF^j is the available demand fraction after the failure of pipe j, and ADF^{jk} is the available demand fraction after the simultaneous failure of pipes j and k. Furthermore, MA stands for the mechanical availability of the system that can be calculated as:

$$MA = \prod_{j=1}^{n} MA_j \tag{6.13}$$

which is the probability that all n pipes are operational. The mechanical availability of pipe j is defined as:

$$MA_j = \frac{MTTF_j}{MTTF_j + MTTR_j} \tag{6.14}$$

where $MTTR_j$ is the *mean time to repair* and $MTTF_j$ is the *mean time to failure* of this pipe, which is given by:

$$MTTF_j = \frac{1}{\lambda_j L_j} \qquad (6.15)$$

Furthermore, in Equation 6.12, the probability of a failure of the jth pipe and all the other components remaining fully functional is given as:

$$u_j = MA \frac{MU_j}{MA_j} \qquad (6.16)$$

For simultaneous failure of pipes j and k:

$$u_{jk} = MA \frac{MU_j}{MA_j} \frac{MU_k}{MA_k} \qquad (6.17)$$

In both equations, MU stands for the mechanical unavailability that equals $1-MA$.

Applying the same procedure to any available demand scenario makes it possible to compile the reliability and availability curves for a 24-hour period, which actually resembles the shape of the expected pressure patterns, or the diurnal patterns turned upside down, as is shown in Figure 6.17.

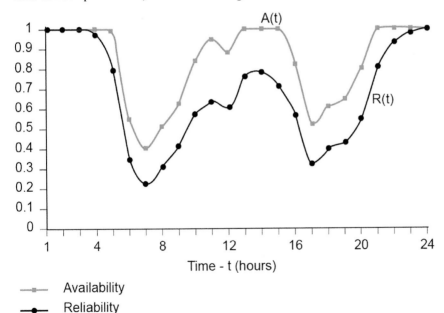

Figure 6.17 Hydraulic reliability and availability curves for diurnal demand patterns

Assessment of network resilience

Todini (2000) describes network resilience using the concept of power balance between (1) the sources of supply, (2) the dissipated amount due to the network hydraulic losses, and (3) the remainder at the discharge points. It is valid in all situations that:

$$\sum_{s=1}^{l} Q_s H_s + \sum_{p=1}^{k} Q_p h_p = \sum_{i=1}^{n} Q_i H_i + \sum_{j=1}^{m} Q_j h_{f,j} + \sum_{v=1}^{t} Q_v h_{m.v} \tag{6.18}$$

$H_{s/i}$ in Equation 6.18 indicates the piezometric heads at l sources (which includes all the reservoirs and tanks that supply the network), and n nodes (which includes all the junctions with discharge, and tanks that are supplied from the network), respectively. The values of $h_{p/f/m}$ include the heads of k pumps, the pipe friction losses of m pipes and the minor losses of t major valves. In all cases, $Q_{s/p/i/j/v}$ stands for corresponding flows supplying the network (s), being conveyed through it (p, j and v), or withdrawn from it (i). Todini's resilience index, I_r, is a measurement of the network's vulnerability of letting some nodes without service in the occurrence of failure, expressed as:

$$I_r = 1 - \frac{P^*_{int}}{P^*_{max}} \tag{6.19}$$

P^*_{int} in Equation 6.19 is the amount of power dissipated in the network to satisfy the total demand, $\Sigma\, Q_i$, under regular operation i.e. at the minimum required pressure, and P^*_{max} is the maximum power that would be dissipated internally during a pipe failure in order to satisfy the constraints of the demand and pressure in the nodes. By substituting Equation 6.18 into 6.19, the I_r can be expressed as:

$$I_r = \frac{\displaystyle\sum_{i=1}^{n} Q_i \left(H_i - H_i^*\right)}{\displaystyle\sum_{s=1}^{l} Q_s H_s + \sum_{p=1}^{k} Q_p h_p - \sum_{i=1}^{n} Q_i H_i^* - \sum_{v=1}^{t} Q_v h_{m.v}} \tag{6.20}$$

where H^* is the minimum piezometric head required to satisfy the demand in node i, equal to the sum of the nodal elevation and the PDD threshold pressure.

Prasad and Park (2004) proposed a more conservative value of the index by introducing weighting of the nodal power, based on possibly larger discrepancies in diameters of connecting pipes. The multiplier c_i is expressed as:

$$c_i = \frac{\displaystyle\sum_{j=1}^{m,i} D_j}{m_i \max\{D_j\}} \tag{6.21}$$

where D_j are the diameters of m pipes connected in node i. The corresponding network resilience, I_n, will be then calculated as in Equation 6.22.

$$I_n = \frac{\displaystyle\sum_{i=1}^{n} c_i Q_i \left(H_i - H_i^*\right)}{\displaystyle\sum_{s=1}^{l} Q_s H_s + \sum_{p=1}^{k} Q_p h_p - \sum_{i=1}^{n} Q_i H_i^* - \sum_{v=1}^{t} Q_v h_{m.v}} \tag{6.22}$$

Hydraulic reliability diagram

The averaged reliability/resilience indices describing one network in its entity say little to nothing about the area affected by the failure i.e. *the impact coverage*, nor about the extent of the failure i.e. *the impact intensity*. The same values of these indices can in theory be the result of a large network area affected to a lesser extent, or a small network portion that is affected severely. Furthermore, little can be grasped about the buffer (i.e. the redundancy) of the network available for conveyance or the delivery of regular flows at the time of failure. Consequently, no viable investment decision can be made purely based on the proposed increase of the index as long as it is not clear what it adds in practical terms i.e. to the improvement of service levels in irregular situations.

To address these deficiencies, Trifunović (2012) proposes the concept of the *Hydraulic Reliability Diagram* (HRD) as a visual representation of the network reliability. The departure point in this consideration is the fact that pipes conveying larger flows also create more severe impact on the demand in the case of their failure. Thus, simulating a single failure of all pipes in the network in a PDD mode gives an opportunity to generate an HRD by plotting the dots indicating each pipe with a pair of values: the fraction of flow under the regular/target demand scenario (Y axis) and the loss of demand, 1-ADF, when the pipe is isolated due to failure/repair (X axis). An example of the diagram for a very simple network composed of one source and seven pipes connected in two loops is shown in Figure 6.18.

In general, for sufficient head at the source(s), all the dots will be located left of the diagonal in the graph. Those potentially located on the right side will have a bigger loss in demand than the flow they convey in regular operation, which suggests that the source head(s) are not sufficient to supply the target demand, even without any pipe failure. Thus, when the system becomes under-designed for the targeted demand, the dots will gradually move to the right side of the diagram, while increasing the redundancy will make them migrate more to the left. Apart from the buffer indication, the position of the dots in the diagram also says something about the network connectivity; in principle, the dots closer to the top indicate lower connectivity while those positioned lower in the graph originate from more connected networks. Finally, a 'special' case of all the dots laying on the diagonal indicates a branched network configuration.

Assuming further that the dots positioned on the Y axis of HRD reflect 100 % of the network buffer i.e. an ADF index equal to one, while the dots positioned on the diagonal reflect no buffer in the pipes, all the other footprints in between these two extremes characterise a particular reliability level that can be correlated to a unique index derived directly from the HRD. This empirical index has been named the *Network Buffer Index* (NBI) and is determined using the analogy between the network buffer and actual loss of demand shown in Figure 6.18. By adding weighting proportional to the pipe flows under regular supply conditions, the NBI will be calculated as in Equation 6.23.

Network Buffer Index

$$NBI = \sum_{j=1}^{m} \frac{Q_j/Q_{tot} - (1-ADF_j)}{Q_j/Q_{tot}} \cdot \frac{Q_j}{\sum_{j=1}^{m} Q_j} = \sum_{j=1}^{m} \left(\frac{Q_j}{Q_{tot}} + ADF_j - 1 \right) \frac{Q_{tot}}{\sum_{j=1}^{m} Q_j} \quad (6.23)$$

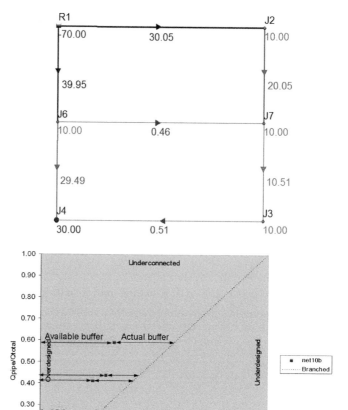

Figure 6.18 Determination of the Network Buffer Index

Source: Trifunović, 2012.

With additional derivation, Equation 6.23 transforms into a simpler form:

$$NBI=1-\sum_{j=1}^{m}(1-ADF_{j})\frac{Q_{tot}}{\sum_{j=1}^{m}Q_{j}}=1-\frac{\sum_{j=1}^{m}(Q_{tot}-Q_{tot,j})}{\sum_{j=1}^{m}Q_{j}}=1-\frac{\sum_{j=1}^{m}\Delta Q_{tot,j}}{\sum_{j=1}^{m}Q_{j}} \quad (6.24)$$

where $Q_{tot,j}$ is the total demand in the network that is available after the failure of pipe j and $\Delta Q_{tot,j}$ is the corresponding loss of demand.

For configurations of serial and branched networks, Equation 6.24 yields the NBI value of 0 (meaning that any pipe failure will cause the loss of the entire downstream demand), while in the case of $1-ADF_{j}=0$ for all the pipes, the NBI will equal 1 (not a single pipe failure will cause the loss of demand because in all the cases the pressure remains above the threshold in the entire network). Finally, the NBI values can also be negative which indicates problems with the source heads, insufficient boosting, high friction losses, or a combination of these causes.

As an example, the calculated NBI value for the network in Figure 6.18 would be 0.499 (Trifunović, 2012).

To demonstrate the concept in a real case network, the HRD and the corresponding NBI have been determined for a portion of the water distribution network for the city of Amsterdam, NL, as shown in Figure 6.19.

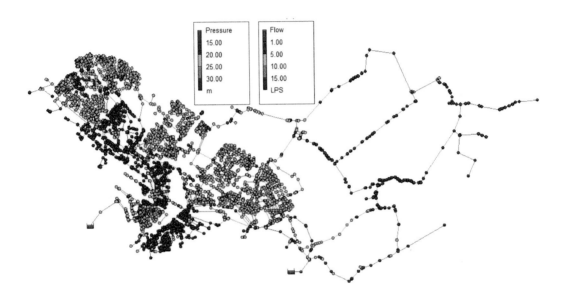

Figure 6.19 Network for Amsterdam North

Courtesy: Waternet, Amsterdam.

The constructed EPANET model consists of 4552 nodes and 5044 pipes and the connection to the rest of the network is simulated by two dummy reservoirs at fixed heads of 25 msl (on the left side) and 21 msl (on the right side). The colours in the figure depict the hydraulic performance of the network for a snapshot simulation of when demand is 468 l/s. Most of the pressures under normal operating conditions fall within the range of 15 to 25 mwc, with two nodes below the PDD threshold pressure of 15 mwc (13.83 and 12.30 mwc, which is the lowest pressure in the network, in the node connected to the dummy reservoir on the right-hand side). The pipe flows are relatively low; a considerable number of the pipes have a flow rate below 1.0 l/s, suggesting a network that is relatively oversized i.e. over-connected. This is confirmed by looking at the corresponding HRD shown in Figure 6.20.

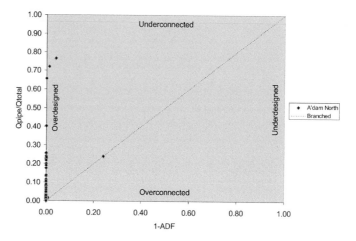

Figure 6.20 HRD for Amsterdam North (Q_{tot} = 467.83 l/s, PDD threshold = 15 mwc)

Therefore, the failure of the majority of the pipes shows no loss of demand whatsoever. The NBI for this network has been calculated at 0.970. The average ADF for the network equals 0.9996, which indicates a highly reliable network but in fact the result is influenced by the huge amount of pipes conveying mostly small flows in a network of nearly 500 loops. At the same time, the calculation of resilience indices, I_r, and I_n, will yield values of 0.605 and 0.560, respectively, which is not an order of value immediately suggesting a high reliability.

To further explore the future network capacity, the HRD has been plotted for three levels of demand increase: 32 %, 73 % and 98 %, which is shown in Figure 6.21. In the most extreme growth scenario, NBI = 0.479 while I_r = 0.286 and I_n = 0.263. Yet, the ADF for the network is still very high (0.9938), which can be explained by zooming in on the area close to the origin of the diagram, which is shown in Figure 6.22.

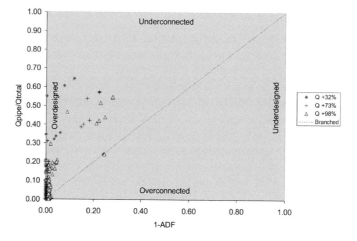

Figure 6.21 HRD for Amsterdam North for increased level of demand

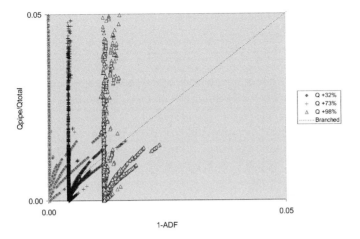

Figure 6.22 Zoom-in of Figure 6.21

This figure shows that some pipes will not be able to deliver the demand due to the pressure dropping below the threshold, but that drop is relatively low and the pipes carry very small flows in a highly looped network, therefore with an insignificant negative effect on the value of ADF.

Technical measures for reliability improvement

A single transportation pipe has practically no reliability as any burst will likely result in a severe drop in supply and pressures; during the repair all the downstream users will have to be temporarily disconnected (Figure 6.23). A burst in the case of parallel pipes causes a flow reduction dependant on the capacity of the pipe/pumping station remaining in operation, say 50 %.

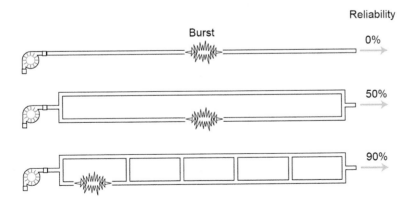

Figure 6.23 Reliability assessment.

Further improvement of the reliability will be achieved by introducing the following technical provisions:

- parallel pipes, pipes in loops, cross connections,
- pump operation with more units,

- alternative source of water,
- alternative power supply,
- proper valve locations,
- pumping stations and storage connected with more than one pipe to the system,
- reservoirs with a larger emergency volume and/or more compartments,
- bypass pipes around the pumping stations and storage, etc.

A few examples of possible cross-connections are shown in Figure 6.24. In long transmission lines, these are usually constructed every 4-5 km, in distribution mains every 300-500 m and in rural areas every 1-2 km.

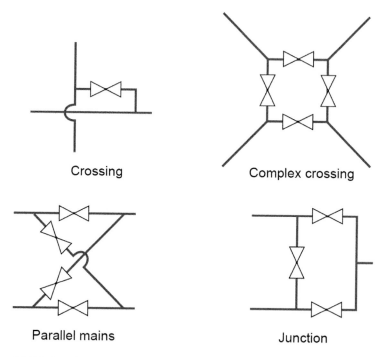

Crossing   Complex crossing

Parallel mains   Junction

Figure 6.24 Technical provisions for improvement of reliability

Source: Van der Zwan and Blokland, 1989.

Setting standards in technical measures that can improve the network reliability is rather difficult due to the variety of situations and consequences that can occur. Nevertheless, some guidelines may be formulated if there are more serious failures. For instance, the Dutch Waterworks Association (VEWIN) proposes 75 % of the maximum daily quantity as an acceptable minimum supply in irregular situations. This should be applicable for a district area of ±2000 connections. Within such an area, valves should be planned to isolate smaller sections of 10-150 connections, when necessary.

Pipe criticality analysis

The classification of pipes based on the flows they convey suggests that these pipes have different levels of significance for the hydraulic operation of water distribution networks. This so-called *hydraulic significance* differs from the significance that considers the structural condition of the pipe. The combination of both aspects defines the *pipe criticality* as an indicator of pipe rehabilitation and eventually a replacement.

Hydraulic significance

Zhou (2018) points out the lack of uniform interpretation of pipe criticality; for some researchers the term means the consequence i.e. the cost of the pipe failure while others also include the probability of the failure. Therefore, pipe criticality also lacks a universally accepted measure. Zhou further cites the significance index (SI) of Arulraj and Rao (1995) shown in Equation 6.25:

$$SI_j = \frac{Q_j L_j}{C_{hw,j} D_j} \tag{6.25}$$

where SI_j stands for the significance index of pipe j, having the length L_j, the diameter D_j and the Hazen-Williams factor $C_{hw,j}$, and conveying the flow Q_j. This simplified and straightforward expression brings the physical properties and condition of the pipe into logical correlation with the hydraulic parameters, by giving higher index values for older pipes of longer length and smaller diameter i.e. the higher burst rate, conveying comparatively high flows. Nevertheless, the value of the index can be distorted by applying the wrong units, and also similar index values may depict a quite different combination of the parameter values; e.g. a short, corroded, large diameter pipe with a high flow may be equally as significant as a longer, smooth, small diameter pipe conveying a low flow. Thus, as is the case with various resilience indices, not much can be concluded in practical terms based on the SI values except that one pipe is more/less significant than another (in theory).

More recent literature proposes a wider range of parameters influencing pipe degradation but these require often-lacking longer and accurate records. Therefore, pipe criticality analysis still mostly relies on network hydraulic reliability that can be optimised by powerful computer models. These models also form the basis of various multiple criteria decision-making tools for whole-life cost analysis which are increasingly used in the asset management of water transport and distribution networks.

6.1.3   Non-revenue water and leakage

The charged water quantity will always be less than the supplied volume. Moreover, the volume of water actually consumed is also smaller than the supplied amount, be it charged or not. The difference in the first case refers to *non-revenue water* (NRW) while the second case represents water leakage, which is often a major component of NRW. Other important factors can be faulty water meters, illegal connections, the poor education of consumers, etc.

Non-revenue water

In older references, the difference between the supplied and authorised consumption used to be defined as *unaccounted-for water* (UFW), the term mostly referring to the current definition of water losses but suffering from diverse interpretations and therefore slowly being phased out from water-loss management vocabulary. In fact, the glossary of terms in this field is under continuous development. Most of it in current use has its departure point in the standard water balance terminology proposed by IWA (2000), which is shown in Table 6.3.

Table 6.3 Structure of a water supply system input volume

Authorised consumption	Billed authorised consumption	Billed metered consumption (including exported water)	Revenue water
		Billed unmetered consumption	
	Unbilled authorised consumption	Unbilled metered consumption	
		Unbilled unmetered consumption	
Water losses (UFW in old references)	Apparent losses (commercial losses)	Unauthorised consumption	Non-revenue water
		Metering inaccuracies	
	Real losses (physical losses)	Leakage in transmission and distribution lines	
		Leakage and overflows at storage tanks	
		Leakage in service connections up to customer meters	

Source: IWA, 2000.

The total system input volume shared between the authorised consumption and the water losses assumes '*the volume of treated water input to that part of the water supply system to which the water balance calculation relates*'.[1] This may include water from external bulk suppliers but does not include the losses prior to (raw water transport) or during the water treatment process. It is also to be noted that this total figure can be affected by the inaccuracy of bulk water meters.

Authorised consumption measures the volume of registered customers, regardless of whether it is billed or not, and whether it is metered. Examples of unbilled authorised consumption is the water used for flushing of distribution pipes, washing streets, firefighting, public taps and fountains, etc. The volume used for these purposes is usually marginal compared to the total water supplied, which also makes the difference between NRW and water losses (former UFW) relatively small in many water supply systems.

The proportion of the components contributing to the water balance says something about the specific operational problems in the system. An example of the water balance in 2008 for the city of Blantyre (Malawi) is given in Table 6.4; it shows a population of approximately 660,000 and a total annual system input of 29,898,713 m³ registered (Kafodya, 2010).

1 The quoted definitions are based on the work of the IWA Water Loss Task Force, established in 1997 to develop standard terminology for water loss balance and water loss performance indicators.

Table 6.4 Water balance in Blantyre, Malawi in 2008, accuracy ± 4 %

		Billed metered consumption 14,949,450 m³ (50.0 %)	Revenue water 14,949,450 m³ (50.0 %)
Authorised consumption 15,044,289 m³ (50.32 %)	Billed authorised consumption 14,949,450 m³ (50.0 %)	Billed unmetered consumption 0 m³ (0.0 %)	
	Unbilled authorised consumption 94,839 m³ (0.32 %)	Unbilled metered consumption 88,893 m³ (0.3 %)	
		Unbilled unmetered consumption 5,946 m³ (0.02 %)	
Water losses 14,854,424 m³ (49.68 %)	Apparent losses 1,931,321 m³ (6.46 %)	Unauthorised consumption 932,425 m³ (3.11 %)	Non-revenue water 14,949,263 m³ (50.0 %)
		Metering inaccuracies 716,137 m³ (2.4 %)	
		Data handling and billing errors 282,759 m³ (0.95 %)	
	Real losses 12,923,103 m³ (43.22 %)	Leakage in transmission and distribution lines and service connections 12,923,103 m³ (43.22 %)	

Source: Kafodya, 2010.

As can be seen, minor adaptations of the IWA table are possible in some specific instances: limited data to distinguish between some components, as is the case with real losses, or actually some additional problem, as is the case with poor data handling and billing errors made by the local water company. It is noteworthy that in the case of Blantyre, exactly half of the system input can be classified as non-revenue water, with high leakage levels of 43.2 %. This is unfortunately not a surprising picture, given the average state of water distribution assets in many developing countries.

In addition to leakage, a significant impact on water losses in developing countries commonly originates from the malfunctioning of water meters, their inaccurate or irregular reading, or from illegal connections. All these contribute to the high NRW levels, as shown in Table 6.5 (Thiadens, 1996).

Table 6.5 Water losses in developing countries

	Water-loss components	Bandung (Indonesia)	Chonburi (Thailand)	Petaling Jaya (Malaysia)
Real losses	Trunk mains, distribution system	21 %	2 %	2 %
	Service connections	10 %	34 %	17 %
Apparent losses	Illegal connections	6 %	2 %	2 %
	Meter under-registering and billing	6 %	8 %	15 %
	TOTAL Water loss	43 %	46 %	36 %

Adapted from: Thiadens, 1996.

Similar figures in the developed world are much lower. Typically, the NRW levels in Western Europe are between 5 and 15 %. A breakdown of the volume distributed in the city of Nuremberg in Germany is given in Table 6.6 (Hirner, 1997).

Table 6.6 The water balance in Nuremberg, Germany

Authorised consumption 91.3 %	Charged water - 84.8 %	
	Bulk supply water - 6.2 %	
	Public use (park, fountains, etc.) - 0.3 %	
Water losses 8.7 %	Apparent losses - 3.5 %	Unmetered usage - 0.5 %
		Own waterworks usage - 1 %
		Meter errors - 2 %
	Real losses - 5.2 %	Pipe breaks - 3.5 %
		Domestic connections - 1 %

Adapted from: Hirner, 1997.

Background leakage

Most leaks occur at pipe joints and service connections. These result in a relatively small water loss per leak, but ultimately a large total loss due to the numerous locations where leakages take place. For this reason, water companies need to be equally concerned by the losses in the distribution system that may pass unattended for a long period of time. This is known as *background leakage* which may be quite substantial while at the same time difficult to survey and repair. On the other hand, pipe bursts, although resulting in more severe water losses, are sudden events of failure that can be easily detected on the monitoring devices existing in the network, showing an instant increase in demand in parallel with a pressure drop. Moreover, in many cases pipe bursts will result in significant amounts of water appearing on the surface, and in extreme cases substantial damage to the surface itself. An example of a suspected pipe burst is shown in Figure 6.25, by the flow and pressure measurements in two sensors placed within the network: P6 and P7 (Sophocleous, 2018). The detection and repair in this case still took several days because the leak was not visible on the surface and therefore pinpointing it accurately was difficult.

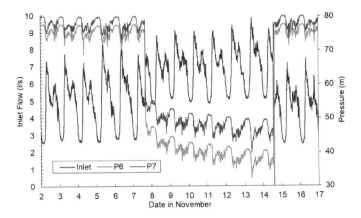

Figure 6.25 Pipe burst event observed by monitoring equipment

Source: Sophocleous, 2018.

In principle, the more severe the burst, the easier the detection and the shorter the repair times. In the case of delayed location and repair, smaller leaks actually create bigger headaches, as can be seen in Figure 6.26 (Farley, 2001). AZNP in the figure stands for *Average Zone Night Pressure*.

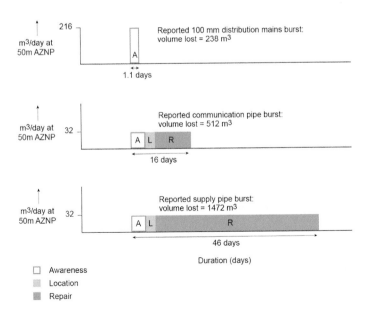

Figure 6.26 Comparison of water lost from different types of leakage

Source: Farley, 2001.

For the sake of simplicity, NRW and water losses (former UFW) are still commonly expressed as a percentage of total system input volume, which can also be seen in the examples shown in figures 6.27 and 6.28. The statistics in these two figures show very high percentages for many cities in developing countries where water is often scarce. Nevertheless, the gross UFW percentage does not necessarily coincide with the losses spread over the length of the network, as the example of records for the cities in Figure 6.29 shows. It is quite possible that higher loss percentages lead to lower losses per kilometre of the network length, as is the case of e.g. Sao Paolo (Brazil), while in the case of Bogota (Colombia) the situation is exactly the opposite. The size/length of the network is obviously different in these two cases.

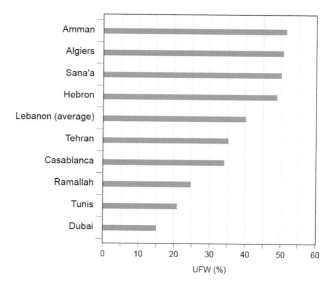

Figure 6.27 UFW in various countries in the MENA Region
Source: World Bank, 2000.

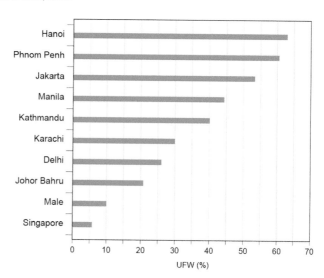

Figure 6.28 UFW in various Asian cities
Source: ADB, 1997.

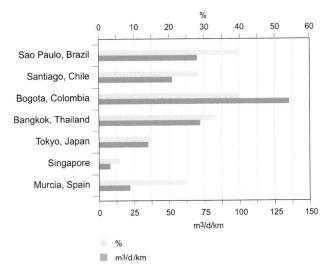

Figure 6.29 Water loss expressed in two different ways

Source: World Bank, 1996.

Looking at it from an economic perspective, NRW will be better expressed as a percentage of the total operating cost of the network rather than a percentage of the system input volume which in a way favours utilities with high consumption and pressures rather than those with intermittent supply at lower pressures. When the network is fully pressurized, NRW and water losses can alternatively be expressed in *litres/day/connection*. Furthermore, apparent (commercial) losses can also be expressed as a percentage of the authorized consumption while real (physical) losses should also be determined by taking the average pressure into consideration.

IWA has proposed the *Infrastructure Leakage Index* (ILI), which is increasingly being used as the performance indicator reflecting the leakage management performance of a water utility. This index is represented in Equation 6.26 as:

Infrastructure Leakage Index

$$ILI = \frac{CARL}{UARL} = \frac{CAPL}{MAAPL} \tag{6.26}$$

where the difference between the parameters in the numerator and denominator, respectively is only in the terminology, which may vary in the literature. CARL (*Current Annual Real Losses*) = CAPL (*Current Annual Physical Losses*), while UARL (*Unavoidable Annual Real Losses*) = MAAPL (*Minimum Achievable Annual Physical Losses*). Depending on the consistency of the selected units of volume over time (per connection), ILI will have no dimension, which increases its applicability in comparing various distribution networks.

The UARL parameter is based on the fact that zero leakage does not actually exist. There will always be some minimum leakage levels depending on the system conditions and operating pressures. Running a statistical analysis of multiple distribution networks in various countries, IWA proposed the following empirical equation for UARL:

$$UARL = \left(A \times L_m + B \times N_c + C \times L_p \right) \times p \tag{6.27}$$

where L_m stands for the total length of distribution mains in the observed area, N_c is the total number of service connections, L_p is the total length of service pipes between the distribution main and the customer meters, and p is the average operating pressure. Coefficients A, B, and C will depend on the units applied. Equation 6.27 is based on the work of Lambert *et al.* (1999) who made a statistical analysis of 27 distribution networks from 20 different countries and came up with values for the coefficients of A = 18, B = 0.8 and C = 25, only for L_m and L_p expressed in km and p in mwc, which gives UARL in litres per day. Thus:

$$UARL = \left(18 \times L_m + 0.8 \times N_c + 25 \times L_p\right) \times p \qquad (6.28)$$

Regardless of whether the sample of networks used to arrive at these values was sufficiently extensive and diverse to confirm its general applicability, and also the linear leakage/pressure relation is often not uniform in practice, Equation 6.28 is widely used in calculating UARL for benchmarking of leakage levels in distribution networks. When UARL is expressed in litres/day/connection, Equation 6.28 must be modified as follows:

$$UARL = \left(18 \times \frac{L_m}{N_c} + 0.8 + 25 \times \frac{L_p}{N_c}\right) \times p = \left(\frac{18}{D_c} + 0.8 + 25 \times L_{p,avg}\right) \times p \quad (6.29)$$

where $D_c = N_c/L_m$ represents the average density of service connections per km of distribution mains, while $L_{p,avg} = L_p/N_c$ is the average length of service pipe between the distribution mains and the customer meters (in km). In many cases, the building i.e. the customer meters are located close to the edge of the street making $L_{p,avg}$ in Equation 6.29 comparatively small ($L_p \ll L_m$ in Equation 6.28), in which case $L_{p,avg}$ and L_p can be neglected. The range of UARL (MAAPL) values in this case is shown in Table 6.7.

Table 6.7 UARL (MAAPL) in l/d/conn for customer meters located at the edge of the street

Density of connections per km of mains	Average operating pressure, p (mwc)				
	20	40	60	80	100
20	34	68	112	146	170
40	25	50	75	100	125
60	22	44	66	88	110
80	21	41	62	82	103
100	20	39	59	78	98

Equations 6.28 and 6.29 characterise networks in reasonably good condition, which are fully pressurised and where regular leakage control takes place. In the

case of intermittency, the UARL (MAAPL) values need to be corrected proportionally to the number of actual supply hours; hence:

$$UARL_{iws} = UARL \times \frac{T_{iws}}{24} \tag{6.30}$$

where T_{iws} represents the actual hours of supply per day in the intermittent water supply regime. Alternatively, the value for CARL can be corrected in the same way:

$$CARL_{iws} = CARL \times \frac{24}{T_{iws}} \tag{6.31}$$

leading to the same ILI_{iws} value:

$$ILI_{iws} = \frac{CARL}{UARL_{iws}} = \frac{CARL_{iws}}{UARL} \tag{6.32}$$

The CARL (CAPL) values are usually derived from water balance analysis, as presented in Table 6.4. Taking the real annual loss indicated there at 12,923,103 m³, and the total number of service connections of 45,323, would lead to CARL values of 12,923,103×1000/365/45,323 = 781.2 l/d/conn. With the total length of distribution mains of 1354 km (the average density of connections of 33.5 per km), the CARL can also be presented as 12,923,103×1000/365/1354 = 26,149 l/d/km. Assuming further the operating pressure of p = 65 mwc, and L_p = 906 km ($L_{p,avg}$ = 20 m), the UARL value would be calculated at 119.4 l/d/conn leading to the ILI of 781.2/119.4 = 6.5. More detailed results for the districts of Blantyre are shown in Table 6.8 (Kafodya, 2010).

Table 6.8 ILI index in the districts of Blantyre

Zone	Mains (km)	No. connections	CARL (m³/y)	UARL (m³/y)	ILI
Kabula	240	14,094	4,018,609	537,180	7.5
Soche	188	12,307	3,509,129	459,863	7.6
Midima	186	7,833	2,233,497	321,026	7.0
Mudi	740	11,089	3,161,868	658,032	4.8
TOTAL	1354	45,323	12,923,103	1,976,100	6.5

Source: Kafodya, 2010.

The lower ILI values depict better performing networks and the bandwidth, which starts in theory at 1.0[2], may stretch to values in the order of 20, which

---

2 Some literature cites ILI values lower than 1.0 in the case of networks with very low leakage levels. Such is the situation in The Netherlands where the registered range of ILIs is between 0.3 and 1.3 (https://www.leakssuitelibrary.com/netherlands-ilis/). This could be due to inaccurate data input, but could also challenge the accuracy/applicability of the UARL value/formula.

reflects poorly operated networks. IWA recognises four categories of network, but distinguishes between high-income countries (HIC) and low and middle income countries (LIC, MIC, respectively), which can be seen in Table 6.9. Accordingly:

- Category A assumes well-operated systems with low leakage levels where *'further loss reduction may be uneconomic and careful analysis would be needed to identify cost-effective improvements'*.
- Category B assumes systems with *'potential for marked improvements by considering pressure management, better active leakage control practices and better maintenance'*.
- Category C assumes poorly operated systems, which can be tolerated only if water is plentiful and cheap. This would however still be a signal to intensify leakage control practices.
- Category D assumes systems in a critical condition which are suffering from extremely insufficient use of resources and where a leakage control programme would be an imperative.

Table 6.9 The IWA matrix of real (physical) losses

Technical performance category		ILI	Real losses in l/conn./day for systems at average pressure of (mwc)				
			10	20	30	40	50
HIC	A1	< 1.5	n/a	< 25	< 40	< 50	< 60
	A2	1.5 – 2		25 – 50	40 – 75	50 – 100	60 – 125
	B	2 – 4		50 – 100	75 – 150	100 – 200	125 – 250
	C	4 – 8		100 – 200	150 – 300	200 – 400	250 – 500
	D	> 8		> 200	> 300	> 400	> 500
MIC, LIC	A1	< 2	< 25	< 50	< 75	< 100	< 125
	A2	2 – 4	25 – 50	50 – 100	75 – 150	100 – 200	125 – 250
	B	4 – 8	50 – 100	100 – 200	150 – 300	200 – 400	250 – 500
	C	8 – 16	100 – 200	200 – 400	300 – 600	400 – 800	500 – 1000
	D	> 16	> 200	> 400	> 600	> 800	> 1000

Source: IWA, 2018.

The overview of the situation in 71 water utilities from 12 high-income European countries is shown in Figure 6.30 (source: https://www.leakssuitelibrary.com/). This figure shows that a considerable number of these networks need to fight leaks; only a third could be considered satisfactory in this respect (ILI < 2).

As already shown in Figure 6.12, pipes break in various ways for various reasons. Common factors influencing leakage can be split into two groups. The first deals with soil characteristics and the corresponding human activities; the main factors in this group are (see Figure 6.31):

- soil movement and aggressiveness,
- heavy traffic loadings,
- damage due to excavations, and
- damage due to the growing roots of plants.

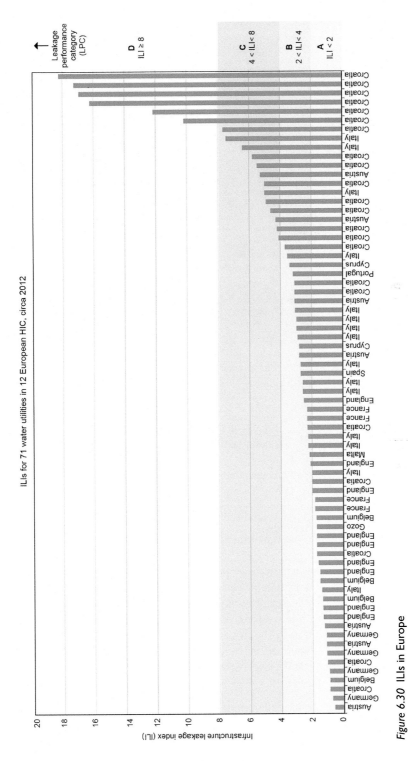

Figure 6.30 ILIs in Europe

Adapted from: https://www.leakssuitelibrary.com/european-ilis/.

Figure 6.31 Damaged steel pipes taken out of operation.

The second group of factors deals with the system components, its construction and operation. Here, the main factors are:

- pipe material, age, corrosion, and defects in production,
- high water pressure in the pipes,
- extreme ambient (winter) temperatures,
- poor quality of joints,
- poor quality of workmanship in pipe laying and jointing.

In many cases the magnitude of the leakage is related to the type of soil. In this respect, the least unfavourable conditions will be met in sandy soils. The soil subsidence expected there can be mitigated by the use of flexible (plastic) pipe materials, and flexible joints. Moreover, an indirect advantage is also that the leaks' visibility is improved by the fact that water normally appears on the surface even during bursts of smaller diameter distribution pipes. Figure 6.32 shows the German experience of acceptable levels of leakage in various types of soils (Weimer, 1992).

Leak detection and location methods

There are four main components of the active real-loss management programme, as shown in Figure 6.33 (Thornton, 2002). The economic level of leakage assumes the total cost of water lost including the cost of the programme, which is assumed to be at a certain minimum. The target of the programme is therefore to bring the real losses down to this economic level, by providing cost-effective measures.

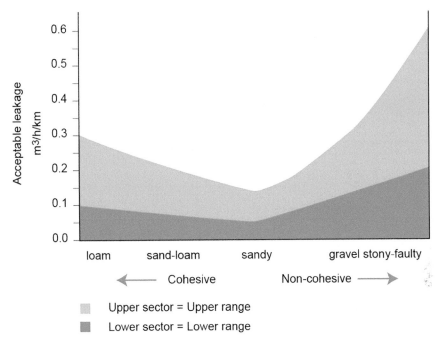

Figure 6.32 Leakage in various soil types

Source: Weimer, 1992.

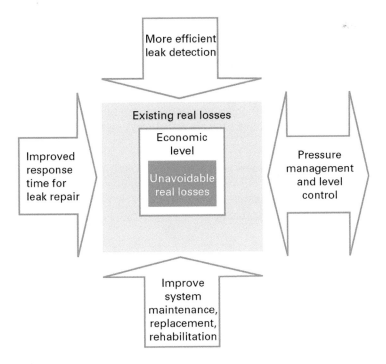

Figure 6.33 Active real-loss management programme

Source: Thornton, 2002.

Global estimates of leakage come from an annual balance of the delivery and metered consumption for the whole network. Bursts of main pipes can be detected by regular flow measurements at supply points (Figure 6.34).

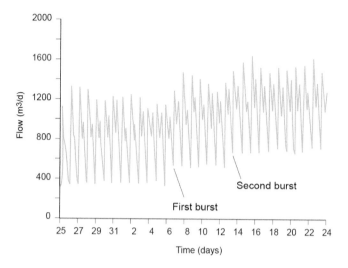

Figure 6.34 Leak detection from the demand monitoring

The above information is not based on specific monitoring of leaks. To enable leak detection, parts of the network need to be more thoroughly inspected. Mea-surements known as *minimum night flow measurements* (MNF) are commonly carried out overnight, when real consumption is at a minimum. A bulk flow meter will be permanently installed with a network of sensors registering abnormal changes in flows and pressures that could originate from leakages. Under normal conditions, the MNF usually occurs between 02:00 and 04:00 hours. Assuming that the (low) consumption in the area is known at that moment, the difference with the MNF can be attributed to the leakage.

Minimum night flow measurements

Alternatively, a van with monitoring equipment can be brought to the location (see Figure 6.35) for conducting what are called *zero consumption measurements*.

Zero consumption measurements

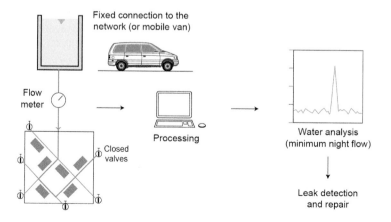

Figure 6.35 District monitoring of night flow.

A smaller area will be isolated from the rest of the system in this case, by closing the district valves and its inflow and outflow will be measured to assess the leakage.

This exercise does not necessarily need to be conducted overnight if it does not cause disruption of normal service.

Thus, leakage measurements can be continuous or periodic, repeated in weekly intervals throughout a period of a few weeks or months. Possible pipe bursts between the periodic measurements should be reflected in a sudden increase in registered (night) demand.

Partitioning of the network for night flow measurements is carried out by forming *district metering areas* (DMAs), as shown in Figure 6.36 (Farley, 2001).

District metering areas

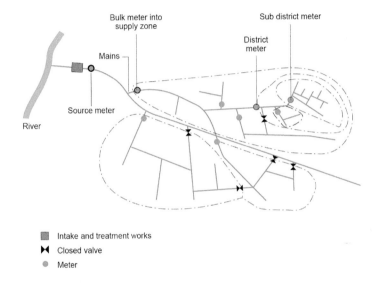

Figure 6.36 District metering areas

Adapted from: Farley, 2001.

The principle of DMA is relatively simple, which is to isolate a zone fed from one supply point. The size of the zone will depend on the complexity of the network and normally includes between a few hundred and a few thousand connections. The topography should also be taken into consideration, as much as possible to avoid large elevation differences within the same DMA. Additional factors to be included are the required economic level of leakage, present distribution of consumption categories, hydraulic conditions, and availability of valves and water meters. The latter will be an important element in the sub-division of DMAs, as shown in Figure 6.37; the process and results of successful implementation of DMAs are illustrated in Figure 6.38 (Farley, 2001).

The average leakage level can also be estimated by measuring pressures in the system, once the relation between the pressure and night flow has been established. Nevertheless, the exact relation between the pressure and leakage is not easy to establish and is very much dependent on the existing pipe materials and their condition. In most cases, the theoretical quadratic relation between the

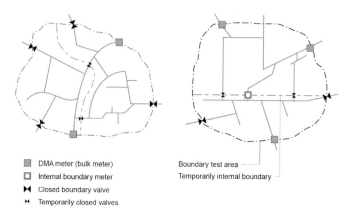

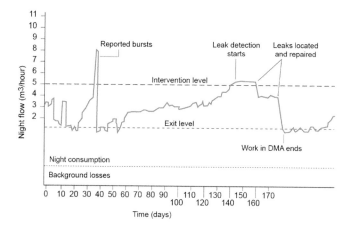

Figure 6.37 Sub-division of a DMA by closing valves (left) and by metering (right)

Source: Farley, 2001.

Figure 6.38 DMA intervention and exit levels

Source: Farley, 2001.

pressure and (leakage) flow does not hold in practice. In the leakage/pressure relation shown by Equation 6.33, L_0/p_0 indicate the initial leakage/pressure while L_1 shows the leakage at revised pressure p_1:

$$L_1 = L_0 \left(\frac{p_1}{p_0} \right)^n \tag{6.33}$$

Figure 6.39 illustrates various possibilities where the selection of exponent n reflects the following situations defined by what is called *fixed and variable area discharge (FAVAD)* (Thornton, 2002):

Fixed and variable area discharge (FAVAD)

- the curve described by $n = 0.5$ indicates a fixed area leakage originating from pin holes appearing mostly in metallic pipes, which are otherwise in fair condition;

- the curve described by n = 1.5 indicates variable area leakage originating from longitudinal breaks, mostly in plastic (PVC) pipes, but also from excessive leakage of joints or high corrosion/background leakage;
- the curve described by n = 1.15 covers average cases.

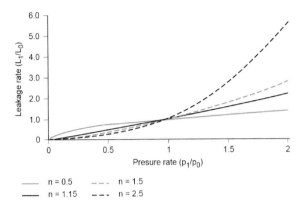

Figure 6.39 Leakage/pressure relation

Adapted from: Thornton, 2002 and Wu *et al.*, 2011.

On a similar note, Wu *et al.* (2011) suggest the following classification based on the n values:

- n = 1.0: not known pipe materials, or a large network with a mixture of materials,
- n = 0.5: fixed area leaks (e.g. holes drilled in thick-walled rigid pipes),
- n = 1.5: variable area leaks whose size depends on pressure,
- n = 2.5: exceptional cases.

Accepting any of these exponents can still lead to inaccurate results due to specific conditions in the network (mixture of materials of different age, progressive increase of leakage over time, undetected background leakage, network renovation, etc.). Assessing these is not easy but if it is possible then it can also lead to a dedicated exponent for a particular network (part). This is also the basis for pressure management as one of the initial strategies in the leakage reduction.

Flow and pressure measurements help to detect leakage levels but do not indicate the exact location of leaks. In the case of severe breaks, water may appear on the surface, but more often leak location techniques have to be applied. The most popular are:

1) the acoustic (sound) method,
2) leak noise correlation, and
3) tracer techniques.

Acoustic detectors rely on sounding directly on a pipe or fitting, or indirectly on the ground surface[3]. The noise generated from the leak is transmitted by a

Acoustic detectors

---

3 Newer applications, mostly in transportation pipes of larger diameters, include the insertion of a ball-shaped acoustic sensor carried with the pipe flow while capturing noises possibly originating from leaks along selected distance which can be up to a few kilometers long. The collected signals are then processed later after the device has been extracted from the pipe. An example of this approach, called *Nautilus System*, can be found at http://www.aganova.es/en/

receiver attached by a stick to an amplifier connected with a stethoscope (Figure 6.40). This method is not always reliable due to the fact that some leaks produce undetectable noise. Moreover, locating the pipe route is much more difficult in the case of plastic and concrete pipes than in the case of metal pipes, as already shown in Table 4.9. Nevertheless, with skilled personnel working under silent (night) conditions, most leaks in metal pipes can be discovered.

Figure 6.40 Leak location by the acoustic method

The equipment is used with devices to locate the pipe route and examples of both acoustic detectors and pipe locators are shown in Figure 6.41.

Acoustic sensors can also be permanently installed in the network in which case they perform leak detection tasks. The network of battery-operated sensors forms a system of noise loggers which all collect the noise signals within their range of sensitivity, usually 150-200 m. These signals can be observed online or processed later. Examples of this equipment are shown in Figure 6.42.

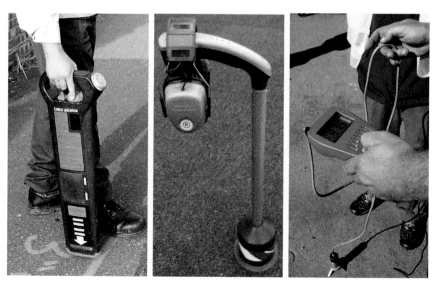

Figure 6.41 Pipe locators (left) and acoustic leak location equipment (centre and right)

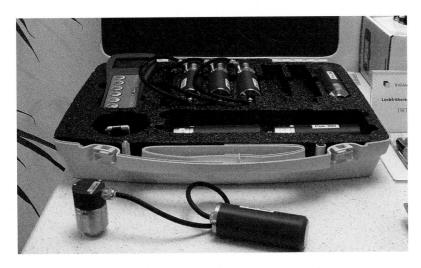

Figure 6.42 Acoustic noise loggers for leak detection

Manufacturer: Wagamet.

Another method based on sound detection is the *correlation method* (figures 6.43 and 6.44). This method uses constant sound propagation in water, which happens at a speed of ±1240 m/s, for detection of leakage. By placing the microphones at the ends of the controlled section, the difference (t) in time required for the leak noise to reach the microphones can be measured by a device called a *correlator*.

Correlation method

Correlator

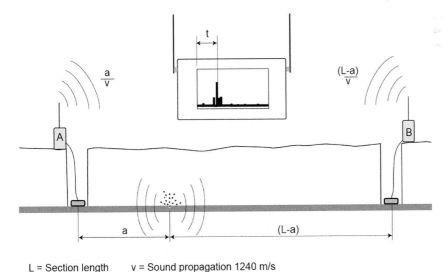

L = Section length   v = Sound propagation 1240 m/s
a = Leak location   t = Time delay in signals from sensors A & B

Figure 6.43 Correlation method - principle

For the known length L of the section, the position of the leak can be calculated. This method is relatively effective in detecting leaks against high levels of background noise and can therefore be applied during the daytime. As in the case of acoustic detectors, it might be less successful in the case of non-metal pipes or if more than one leak exists along the inspected pipe route. Apart from ambient noise and pipe material, other factors generally affecting acoustic leak detection are low water pressure, variations in pipe depth or properties of bedding material, mixture of different soils, and a high groundwater table.

Figure 6.44 Correlation equipment for leak location

Manufacturer: Sewerin.

Tracer method   The *tracer method* involves a gas being pressure-injected into the main that is under inspection; this main will be isolated from the rest of the system for this purpose. As water leaks, its pressure reduces to atmospheric level and the gas comes out of the solution. The presence of the gas is then tested by using the gas detector probe inserted into the holes made along the known pipe route (Figure 6.45).

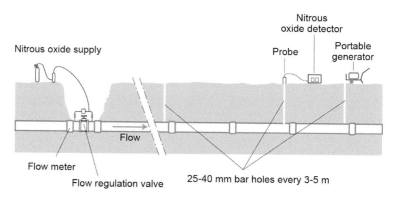

Figure 6.45 The tracer method

A gas frequently used as a tracer is nitrous oxide (N_2O), being non-reactive, non-toxic, odourless and tasteless. It is soluble in water and can be registered in very small concentrations. Other gases can also be used, e.g. sulphur hexafluoride (SF_6). Work with gas tracers in the US includes the inspection of empty pipes that can be pressurised by helium or nitrogen. Depending on the conditions, drilling of probe holes might be avoided in this case (Smith *et al.*, 2000).

The tracer method is generally more expensive than the acoustic method. Its advantage is that it is not dependent on the conditions required for acoustic leak detection. Furthermore, it can be used for locating bursts in empty pipes, which may sometimes be required for emergency reasons.

Water meter under-reading

As is the case with real losses, the apparent loss management program also comprises four major components as shown in Figure 6.46 (Thornton, 2002). The magnitude of the problems puts these into proper perspective. While water theft and the lack of adequately trained human resources in the management of water utilities, mostly in developing countries, may pose significant challenges, the general technical problem related to apparent losses widely spread in many networks is water-meter under-reading. If not regularly maintained and replaced, water meters may register inaccurate amounts of water. On top of this, where the meter reading is not automated, human errors may lead to a significant loss of revenues.

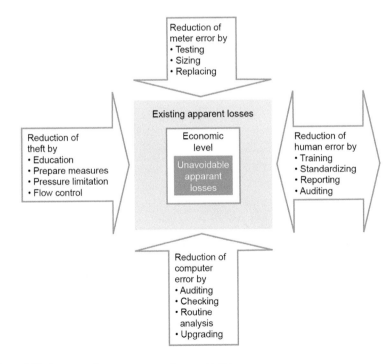

Figure 6.46 Apparent loss management programme

Source: Thornton, 2002.

Regularly maintained meters are capable of registering flows with an accuracy of 95-98 %, with a normally higher error rate in the lower zone of flows. Van Zyl (2011) illustrates South African standards in the example of a particular model of inferential meters, shown in Figure 6.47.

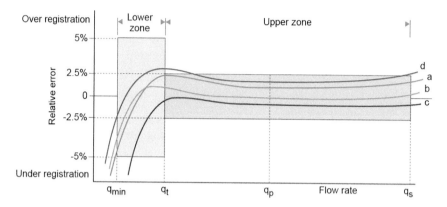

Figure 6.47 Bandwidth accuracy of specific inferential meters

Source: Van Zyl, 2011.

In this figure, the lower zone is defined by the *minimum flow* rate that the meter should be able to register with a maximum permissible error (q_{min}), and the *transitional flow* rate (q_t) at which the maximum permissible error is expected to be reduced. In regular mode, the meter is expected to operate around the *permanent flow* rate (q_p) up to the maximum flow rate, called the *overload flow* rate (q_s). The figure shows the meters a and b complying with the standard, while the meters c and d are not operating satisfactorily.

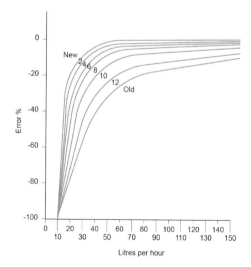

Figure 6.48 Drop in accuracy of water meters in operation

In any case, the accuracy of water meters reduces after a few years in operation (see Figure 6.48). A severe drop in the accuracy of the measurements can occur as soon as five years of operation. This is again particularly evident in the lower range of flow rates. Therefore, the measuring devices have to be regularly checked, and if necessary repaired and re-installed. The choice between the two options will depend on the cost evaluation of each renewal and increased water loss due to malfunctioning (Figure 6.49). The economic period varies between six and 12 years and in exceptional cases longer; water meters are most commonly replaced every eight to ten years.

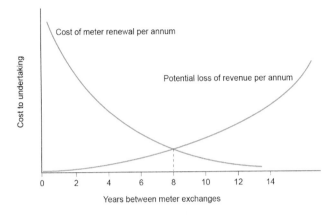

Figure 6.49 Economics of meter exchange periods

The calibration and repair of water meters takes place in hydraulic benches, as shown in Figure 6.50.

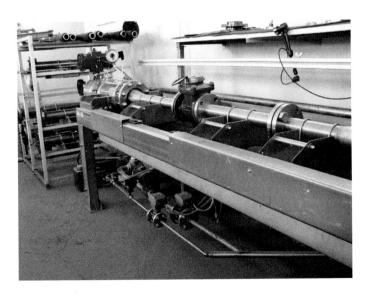

Figure 6.50 Equipment for calibration and repair of water meters

When water meters are installed within the premises, the fact that the real amount of water used is paid for increases consumer awareness. However, this principle is not always economical. It requires additional investment in equipment and personnel involved in installation, maintenance, replacement, reading and administration. The final decision normally depends on local conditions such as: average water demand, network coverage, labour costs, ability of the consumers to pay the bill, etc. Usually, the cost of individual metering is born fully or partly by the consumers.

Organisation of the leak survey programme

The level of leakage will indicate which measures should be taken to rectify it. For leakage levels above ±30 %, a leak survey and metering programme (LSM) is usually justified by the savings obtained. Below that level, an economic study should be carried out taking into consideration the costs of production, distribution and leakage control. The conditions of the water sources are also a relevant element in the final decision. At the very least, by reducing the leakage level, an investment in the system extension can be postponed by several years (Figure 6.51).

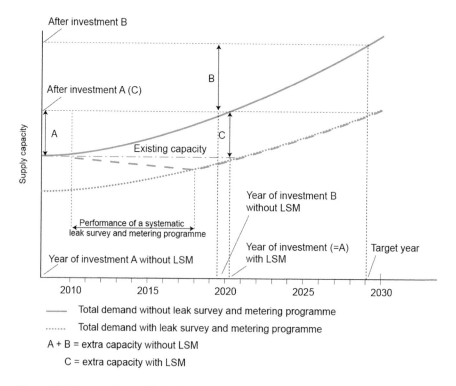

Figure 6.51 Savings obtained by leakage reduction

Figure 6.52 shows the reduction in leakage when applying different levels of control (Brandon, 1984). In the passive method, only major leaks and leaks that are reported by consumers are repaired: this is the reactive approach, in which no systematic effort is made to measure or detect leakage.

Regular sounding is not considered to be the main component of the leakage detection procedure. When it is not selective, regular sounding requires a lot of manpower and sometimes provides unreliable results. Therefore, it is better to apply sounding techniques at a local level where leakage has already been indicated by measurements.

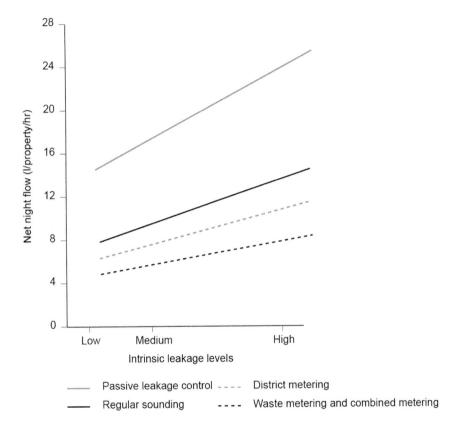

Figure 6.52 Levels of leak control
Source: Brandon, 1984.

The most efficient (and expensive) approach includes prevention and reaction. Once a decision has been taken on systematic leak detection, the activities must be organised carefully. A possible organisational set-up of a leak detection team in a water supply company is shown in Figure 6.53.

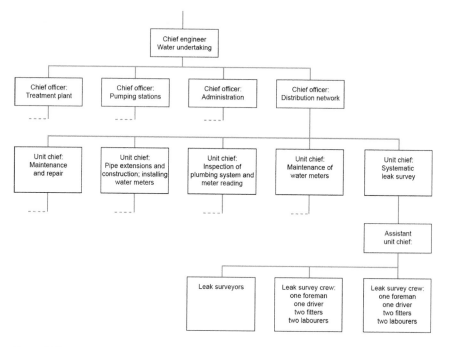

Figure 6.53 Leak detection team set-up

Source: KIWA, 1994.

The main components of the leakage detection programme are:

1)   data collection,
2)   planning,
3)   organisation, and
4)   leak survey and repair.

The data collection comprises information on:

•   pipes: route, material, dimensions, age,
•   measuring devices,
•   valves and fire hydrants, and
•   service connections.

After this has been completed and evaluated, the next step is planning, comprising:

•   preparation of operational charts for specific activities,
•   selection of type and quantity of the equipment needed,
•   planning of a training programme for personnel, and
•   determination of the priorities among the areas surveyed.

The organisational stage consists of:

- selection and training of the personnel, and
- procurement of the equipment.

Finally, the leak survey can commence. The economic viability of the leakage survey programmes is evaluated based on:

- total amount of water loss due to leakage,
- minimum acceptable leakage percentage,
- total cost of water production,
- maximum saving if the minimum acceptable leakage can be achieved,
- investment and labour costs involved in the leakage detection programme,
- maintenance of the system at the minimum-acceptable leakage, and
- the amount of water saved by the programme.

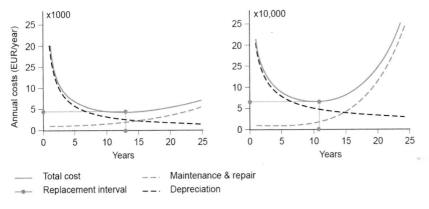

Figure 6.54 Assessment of a pipe rehabilitation programme

Adapted from: Wu et al., 2011.

Various strategies can be applied to reduce the water losses, starting from the pressure management, dealing with water meter inaccuracies, correcting human errors with improved training for better management of the processes, and most radically by a systematic pipe rehabilitation programme including their large-scale replacement. In the latter strategy, the timing of renewal has to be carefully assessed with the failure risks, showing that the large diameter pipes will be more critical and due for earlier replacement than the smaller diameter pipes, as can be seen in Figure 6.54. The link between the reliability assessment and the leakage and asset management is very obvious in this case. An example of the effects of a pipe renovation programme on the leakage reduction programme for the city of Odense in Denmark is shown in Figure 6.55 (Merks et al., 2015). The programme was executed in the period 1995-2013.

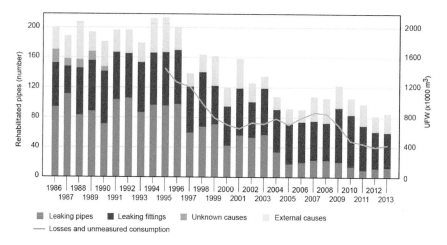

Figure 6.55 Leak reduction in Odense, Denmark, in m³ ×1000

Source: Merks *et al.*, 2015.

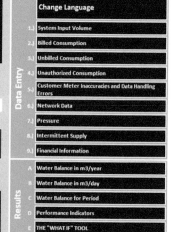

Figure 6.56 Example of water balance software

Source: www.liemberger.cc.

Numerous decision support tools are available nowadays on the Internet to assist in producing a water balance and help selecting appropriate NRW reduction strategies. This includes both the real and apparent losses. One example is WB-EasyCalc, a spreadsheet application in the public domain, developed by Liemberger & Partners (the Miya Group), shown in Figure 6.56.

Problem 6-1 (based on the group work case from IWA, 2018)

Four water utilities perform as shown in the following table:

No.	Water Utility	A	B	C	D
i1	System input volume (SIV), m³/d	50,000	250,000	500,000	1,000,000
i2	Billed metered consumption (BMC), m³/d	15,000	75,000	250,000	800,000
i3	Billed unmetered consumption (BUC), m³/d	25,000	125,000	75,000	40,000
i4	Unbilled metered consumption (UMC), m³/d	1000	2500	3000	8000
i5	Unbilled unmetered consumption (UUC), SIV %	1.0 %	1.5 %	1.0 %	2.0 %
i6	Water theft (illegal connections), BC %	5.0 %	10.0 %	10.0 %	3.0 %
i7	Customer meter under-reading, MC %	5.0 %	8.0 %	10.0 %	3.0 %
i8	Total length of mains (L_m), km	400	2800	6200	12,800
i9	Total number of service connections (N_c)	12,000	112,000	196,000	353,000
i10	Average length of single service connection, m	0	0	5	5
i11	Average pressure (p), mwc	15	15	20	30
i12	Average supply time per day, hrs	12	18	24	24
i13	Average tariff, EUR/m³	0.25	0.35	0.42	1.20
i14	Variable production cost, EUR/m³	0.10	0.15	0.20	0.70
i15	Unsatisfied demand*, m³/d	3000	15,000	100,000	50,000

* - the water which could be sold if physical losses were reduced.

Calculate the main NRW performance indicators and explain the results by proposing the most efficient measures for improvement in each case.

Answer:

The result of calculations are shown in the following table using the water balance parameters from Table 6.3, and Equations 6.28, 6.31 and 6.32, all composed in hydraulic Spreadsheet Lesson 8-12 (Appendix 7).

No.	Water Utility	A	B	C	D
	Volumes				
r1	Metered consumption (MC), m³/d	16,000	77,500	253,000	808,000
r2	Billed consumption (BC), m³/d	40,000	200,000	325,000	840,000
r3	Unbilled consumption (UC), m³/d	1500	6250	8000	28,000
r4	Authorised consumption (AC), m³/d	41,500	206,250	333,000	868,000
r5	Commercial/apparent losses, m³/d	2800	26,200	57,800	49,440
r6	Physical/real losses, m³/d	5700	17,550	109,200	82,560
r7	NRW, m³/d	10,000	50,000	175,000	160,000

(*Continued*)

(Continued)

Water loss performance indicators				
r8 NRW (w.s.p.*), l/d/conn	1667	595	893	453
r9 Physical/real losses (w.s.p.), l/d/conn	950	209	557	234
r10 Commercial/apparent losses, % of a.c.**	6.7 %	12.7 %	17.4 %	5.7 %
r11 UARL (MAAPL) (w.s.p.), m³/d	126	1575	5858	16,708
r12 Infrastructure leakage index (ILI)	45.2	11.1	18.6	4.9

Financial aspects				
r13 Annual value of commercial losses, EUR/y	255,500	3,347,050	8,860,740	21,654,720
r14 Annual value of physical losses, EUR/y***	372,300	2,055,863	16,001,600	30,219,080
r15 Total annual value of NRW, EUR/y	627,800	5,402,913	24,862,340	51,873,800

Percentages of system input volume (SIV)				
r16 Metered consumption (MC), m³/d	32.0 %	31.0 %	50.6 %	80.8 %
r17 Billed consumption (BC), m³/d	80.0 %	80.0 %	65.0 %	84.0 %
r18 Unbilled consumption (UC), m³/d	3.0 %	2.5 %	1.6 %	2.8 %
r19 Authorised consumption (AC), m³/d	83.0 %	82.5 %	66.6 %	86.8 %
r20 Commercial/apparent losses, m³/d	5.6 %	10.5 %	11.6 %	4.9 %
r21 Physical/real losses, m³/d	11.4 %	7.0 %	21.8 %	8.3 %
r22 NRW, m³/d	20.0 %	20.0 %	35.0 %	16.0 %

* - when the system is pressurised
** - percentage of authorised consumption
*** - taking additional water sales into account

The relation between the input data and the results in the above tables is described by the following equations:

$r1 = i2 + i4$
$r2 = i2 + i$
$r3 = i4 + i5 \times i1$
$r4 = r2 + r3$
$r5 = i6 \times r2 + i7 \times r1$
$r6 = i1 - r4 - r5$
$r7 = i1 - r2$
$r8 = r7 \times 24{,}000/i9/i12$
$r9 = r6 \times 24{,}000/i9/i12$
$r10 = r5/r4$
$r11 = (18 \times i8 + 0.8 \times i9 + 25 \times i9 \times i10/1000) \times i11 \times i12/24{,}000$
$r12 = r6/r11$
$r13 = r5 \times i13 \times 365$
$r14 = (i15 \times i13 + (r6 - i15) \times i14) \times 365$
$r15 = r13 + r14$
r16 to r22 = r1/i1 to r7/i1, respectively.

The abbreviation 'w.s.p.' in the case of r8, r9 and r11 means 'when the system is pressurised'; these results include the adjustment related to the hours of intermittent supply. Furthermore, the annual loss of revenue due to physical losses (r14) takes into consideration that not the entire volume of the calculated physical losses can be charged to consumers if the leaks are completely fixed (in theory). If the shortage of the actual demand not satisfied (i15) is less (i15 < r6), the difference (r6 − i15) is the financial loss based only on the production costs (i14). If on the other hand i15 > r6, the total loss would be reduced for the production costs of additional water needed in addition to the recovered physical losses.

Discussion:

Utility A: The smallest system of all four and with the highest degree of intermittency. A high percentage of billed consumption but not much of it is metered. A moderate percentage of illegal use and meter inaccuracy. On average 30 connections per km of mains. The results show 'only' 20 % of NRW but with the highest value when expressed in l/d/conn. ILI is extremely high, pointing to the very poor state of the system. The same conclusions can be drawn for the physical losses which look low in percentage terms but are the highest when expressed in l/d/con and contribute more to the overall loss of revenue than the commercial losses (nearly 60 %). Reducing leaks would therefore be the most obvious short-term strategy.

Utility B: The second smallest but with 5-times larger SIV than A. In the case of billed consumption, similar to A; however, with higher illegal use and meter inaccuracy. 40 conn/km of mains. The results show the same NRW of 20 % as for A but with a much lower value of NRW and physical losses when expressed in l/d/conn. ILI is also significantly lower, still pointing to the unsatisfactory state of the system. The commercial losses are dominant in the overall loss of revenue (nearly 62 %). Improving the accuracy of meters and reducing water theft are therefore the most obvious short-term measures in this case. It is also notable that a 5-times larger SIV causes an 8.6-times larger loss of revenue, which is a consequence of the water tariff and production costs being higher than for A.

Utility C: Twice as high SIV compared to B, and with continuous supply. Significantly lower billed volume but with a higher percentage of metered consumption. Equally high illegal use and meter inaccuracy as in B. 32 conn/km of mains. The results show a significantly higher NRW percentage, also with relatively high NRW and physical losses when expressed in l/d/conn but still lower than in A. ILI is also significantly lower than in A, but still reflects the poor state of the system. The physical losses contribute to the overall loss of revenue of approximately 64 %. Here as well, reducing the leakage is the most obvious short-term strategy. However, the higher percentage of the commercial losses compared to authorised consumption, also deserves an attention.

Utility D: The biggest SIV and with continuous supply. The highest percentage of metered and billed volume, and with a low percentage of illegal use and meter inaccuracy. 28 conn/km of mains. High tariff and water production costs. In essence, a 'Western-style' system operating reasonably well, which is reflected in the lowest NRW percentage of all four, however with NRW and physical losses when expressed in l/d/conn comparable to B. ILI is also the lowest of all four. The physical losses contribute to the overall loss of revenue of approximately 58 %. Due to the high total losses of revenue (high SIV, expensive water), this is the network which needs combined measures; firstly, by reducing the leakage, but also acting on the commercial losses. It is also to be noted that ILI is relatively low due to high average operating network pressure making the UARL value comparatively high. The potential for a NRW reduction programme is also demonstrated by the fact that the revenue from the water is the highest.

Self-study: Spreadsheet Lesson 8-12 (Appendix 7)

6.1.4   Corrosion

Corrosion causes deterioration of material properties due to their reaction with their environment. In water transport and distribution, this process takes place predominantly by attacking pipes and joints, defined as:

- external corrosion, in reaction with the soil, or
- internal corrosion, in reaction with water.

Concerning the materials, two types of corrosion can be distinguished:

1) metallic corrosion, and
2) corrosion of cement-based products.

Metallic corrosion is a chemical reaction caused by the transfer of electrons (Figure 6.57). After a metal ion leaves the pipe surface (anodic site) and enters water, excess electrons migrate through the metal to a cathodic site where they are used by a balancing reaction. Three types of reaction are possible:

1) hydrogen evolution - typical in aggressive waters (with low pH),
2) oxygen reduction - typical in normal waters, and
3) sulphate reduction - typical for anaerobic conditions occurring in soils.

The direct consequence is that the pipe will lose its mass at the anodic site, which will be partly dissolved and partly accumulated at the cathodic site. Practical problems resulting from this are:

- loss of water and pressure due to leakages,
- increased pumping costs due to pipe clogging,
- malfunctioning of appurtenances in the system,
- malfunctioning of indoor installations, and
- the appearance of bad taste, odour and colour in the water.

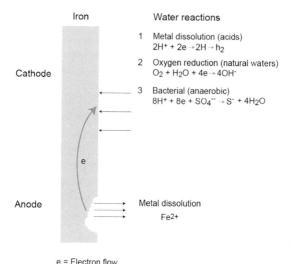

Iron   Water reactions

1   Metal dissolution (acids)
    $2H^+ + 2e \rightarrow 2H \rightarrow h_2$

2   Oxygen reduction (natural waters)
    $O_2 + H_2O + 4e \rightarrow 4OH^-$

Cathode

3   Bacterial (anaerobic)
    $8H^+ + 8e + SO_4^{--} \rightarrow S^- + 4H_2O$

e

Anode   Metal dissolution

                        Fe^{2+}

e = Electron flow

Figure 6.57. Metallic corrosion

Source: Brandon, 1984.

The corrosion of cement-based (or lined) materials is a chemical reaction in which the cement is dissolved due to the leaching of calcium at low pH values (in principle less than 6.0). This can be a problem in the case of cement-lined metal pipes but also with concrete or AC pipes, where the structural strength of the pipe may be lost or a metal (reinforcement) can be exposed to the water enhancing the corrosion process even further.

Corrosion forms

The actual mechanisms of corrosion are usually the interrelation of physical, chemical and biological reactions. Common forms of pipe corrosion are (AWWA, 1986):

Galvanic corrosion - When two different pipe metals are connected, the cathodic site tends to be localised on the less reactive material, and the anodic on the more reactive, causing corrosion known as galvanic corrosion. This kind of corrosion is typical at joints of indoor installations (e.g. copper pipe connected to galvanized iron causes corrosion of the iron). The galvanic corrosion can be particularly severe at elbows. *Galvanic corrosion*

Pitting - This is localized, non-uniform corrosion scarcely detectable before a hole appears. This is a potentially dangerous form of corrosion because even small holes may cause rapid pipe failure. Surface imperfections, scratches or deposits are favourable places for pitting corrosion. *Pitting*

Tuberculation - Tuberculation occurs when pitting corrosion builds up at an anode next to a pit. Tuberculation rarely affects the water quality unless some of the tubercles are broken due to sudden changes in flow. Serious forms of this corrosion lead to a drastic increase of pipe roughness, i.e. reduction in the inner diameter. Hence, this type of corrosion can be suspected by monitoring the hydraulic performance of the pipe. The example in Figure 6.58 shows the effects of tuberculation on the reduction of the cross-section area of CI pipes in the distribution network in Boston. The percentage indicates the remaining area of the pipe cross-sections. *Tuberculation*

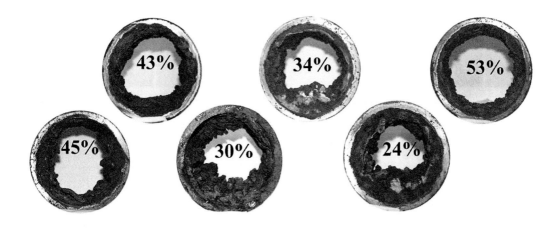

Figure 6.58. Effects of tuberculation on the reduction of the pipe cross-section
Source: Snoeyink, 2002.

Crevice corrosion

Crevice corrosion - A form of localized corrosion usually caused by changes in acidity, oxygen depletion, dissolved ions, etc. Crevices appear at joints or surface deposits.

Erosive corrosion

Erosive corrosion - With this corrosion a protective coating against corrosive attack is mechanically removed due to high velocities, turbulence or sudden changes in flow direction. Pieces of the pipe material can also be removed. This corrosion form is common at sharp bends.

Cavitation corrosion

Cavitation corrosion - This is a type of erosion corrosion caused by the collapse of vapour bubbles (most often at pump impellers, as explained in Section 4.2.4). It occurs at high flow velocities immediately following a constriction or a sudden change in direction.

Biological corrosion

Biological corrosion – The reaction between the pipe material and microorganisms that appear in pipes results in this form of corrosion. Biological corrosion is common in stagnant waters and at dead ends of networks. It is an important factor in the taste and odour problems that develop, but also in the degradation of the material. Control of biological growth is very difficult. It can also appear in pipes where disinfection by chlorine is inefficient.

Corrosion and water composition

Water composition is a key factor that influences internal pipe corrosion. A lot of effort has been put into establishing the quantitative relationship between the water composition and types of corrosion. Factors influencing the internal corrosion of metal pipes according to US experience are listed in Table 6.10.

Specific research, which also covered some non-metal pipes, has been carried out in the Netherlands (Van den Hoven *et al.*, 1988). The following parameters were analysed:

1) pH value,
2) TIC (total inorganic carbon) as the sum of carbon-based elements: (H_2CO_3), (HCO_3^-), (CO_3^{2-}),
3) SI (saturation index) as an indicator of water aggressiveness with respect to calcium carbonate: $(CaCO_3)$, and
4) chloride and sulphate: (Cl^-), (SO_4^{2-}).

The findings per pipe material are listed below.

AC, concrete and cement mortar - Corrosion of cement predominantly caused by acids may result in:

1) disintegration of the inner surface layer of AC pipes and exposure of fibres that might easily be released into the water,
2) increase of pH (especially with water stagnation) and suspended solids, calcium, iron, aluminium and silicates,
3) reduction in pipe strength, and
4) increased energy loss by increased wall roughness.

Present knowledge indicates that corrosion of cement-based materials and the release of asbestos fibres are acceptable when SI > -0.2 (minimal corrosion is obtained for positive SI values), although it is still doubtful whether the saturation index can be assumed to be a sufficient indicator of corrosion in this case.

Ductile iron and steel - Corrosion of these materials is predominantly caused by dissolved oxygen. Other components such as nitrate, hydrogen ions and some chlorine components (where chlorination takes place) can also act as oxidisers. It is also assumed that some bacteria can accelerate the corrosion of ductile iron.

Table 6.10 Factors influencing the internal corrosion of metal pipes

Factor	Effect
pH	Low pH increases the corrosion rate. High pH tends to protect iron pipes but may cause the removal of zinc from brass and damage to copper pipe.
Dissolved oxygen	Increases the rate of many corrosive reactions.
Temperature	Increases the rate of many corrosive reactions. Decreases calcium carbonate solubility. Increases biological activity.
High flow velocity or turbulence	Increases potential for erosive corrosion.
Low flow velocity	Increases potential for crevice and pitting corrosion.
Alkalinity	Helps to form a protective coating of scale and buffer pH changes. High alkalinity increases corrosion of copper, lead and zinc.
Calcium hardness	Helps to form a protective coating of scale but at high concentration may cause tuberculation or excessive scaling.
Chlorine residual	Increases most types of corrosion of metal pipe. Decreases biological corrosion.
Chloride and sulphate	Increases corrosion of iron, copper and galvanised steel. Increases tendency for pitting corrosion.
Hydrogen sulphide	Increases corrosion.
Ammonia	Increases corrosion of copper pipe.
Magnesium	May inhibit precipitation of the calcite for calcium carbonate and favours the formation of the more suitable aragonite form.
Total dissolved solids	Increases water conductivity, which tends to increase the corrosion rate.

Source: Smith *et al.*, 2000.

This all results in the formation of iron deposits on the pipe wall, causing an increased hydraulic resistance and adverse effect on water quality. The release of iron causes the water turbidity to increase and colour to appear (yellow-brown, or green in anaerobic environments). Also, lead absorbs or is built into iron particles, which results in a higher lead intake via drinking water. The following water composition can be recommended:

pH > 7.5
TIC > 0.5 mmol/l
$((Cl^-) + 2(SO_4^{2-})) / TIC < 1$ (known also as the *corrosion index*, CI)   *Corrosion index*

Copper - Two forms of corrosion are typical for copper: uniform corrosion and pitting. Uniform corrosion starts with a chemical reaction between copper and dissolved oxygen (other components such as free chlorine can sometimes also act as oxidisers) and results in the formation of a layer of insoluble copper salts. These layers, usually a green colour, normally stabilise after one or two years without serious effects on the pipe material. However, taste and discolouration may appear due to the copper release.

Pitting corrosion is a more serious problem that can occur within a few months of the installation of the pipe. It appears as a consequence of either poor material quality (carbon particles left behind on the pipe wall during manufacturing), the installation technique (aggressive fluxes and residuals produced during soldering) or water stagnation (particle precipitation). The risk of leakage caused by pitting corrosion can be reduced by increasing the pH, bicarbonate concentration and organic matter; oxygen levels and particle concentration should then decrease. The recommended values are:

TIC > 2 mmol/l, to prevent pitting corrosion and
pH > 0.38TIC + 1.5(SO_4^{2-}) + 5.3, to prevent copper release.

Dezincification

Brass – The zinc present in this alloy dissolves in contact with water (*dezincification*). The part of the pipe affected by this process becomes porous over time resulting in fractures and leaks. The zinc release in the water is insignificant. Experiments have shown that chloride ions affect the rate of dezincification while bicarbonate ions determine the occurrence of the process. High pH values promote the formation of deposits on the pipe wall that can cause clogging. The recommendations are:

pH < 8.3
TIC > 4 mmol/l
(Cl^-) < 40 mg/l

Finally, Van den Hoven *et al.* (1988) give the following general recommendations for optimal water composition in a distribution system:

7.8 + (0.38TIC + 1.5(SO_4^{2-}) + 5.3) < pH < 8.3
TIC > 2 mmol/l
-0.2 < SI < 0.3
CI = ((Cl^-) + 2(SO_4^{2-})) / TIC < 1

The basis for the upper limit of the saturation index (0.3) is the need to inhibit scaling in hot-water apparatus.

The findings of Van den Hoven *et al.* (1988) have been reassessed more recently and updated by Slaats *et al.* (2013), who challenged some of the general recommendations in the previous work, namely:

1. The recommendation on 7.8 + (0.38TIC + 1.5(SO_4^{2-}) + 5.3) < pH, mostly established for copper release in copper pipes, is linked to the maximum-allowed concentration of copper in distribution networks of 3 mg/l, which is no longer the regulatory norm in the Netherlands. Moreover, no statistically significant relation between the calculated and measured values of copper release could be established.
2. The recommendation on pH < 8.3, mostly established for dezincification of brass, was abandoned because no serious record of clogging has been observed in practice, and also the use of brass has reduced since new regulations were introduced in 2011.

3. The basis for the upper limit of SI < 0.3, established to prevent scaling, has also been phased out by introducing newly developed methods for assessment of calcium precipitation by analysing the total hardness, and $TACC_{90}$/$PACC_k$ (mentioned in the table below). Moreover, the upper SI limit may affect the increase in pH needed to reduce the release of copper and lead.

4. The recommendation for corrosion index CI < 1, related to iron release, has been removed from the regulations owing to the conclusion that 'brown' water mostly originates from inadequate water treatment (in the Netherlands).

Instead, Slaats *et al.* (2013) propose the following simplified list of recommendations for optimal water composition, shown in Table 6.11.

Table 6.11 Recommended water composition per pipe material

Material	Parameter	Value	Remarks
Iron	pH	> 7.0	The higher the value, the lower the risk of iron release.
	TIC	> 2 mmol/l	In this range, the formation of a protective layer is possible.
Lead	pH	As high as possible	Demanded only in parts of the network where lead is (still) present.
Copper	pH	> 7.4	In this range, the copper release will be limited.
	TIC	> 2 mmol/l	In this range, prevents the growth of pitting.
Cement-based	SI	> - 0.2	In this range, prevents corrosion attacks on cement-based materials.
Scaling	TH	< 1.8 mmol/l	Stands for total hardness, which describes the concentration of dissolved calcium ions and magnesium ions, expressed in mmol/l, or German degrees (°D), where 1 mmol/l = 5.6 °D. The recommended value is based on customer satisfaction. The presence of calcium and magnesium in drinking water contributes moderately to a healthy diet but too high concentrations increase the use of soap and washing powder and liquids.
	$TACC_{90}$	< 0.6 mmol/l	Translated from Dutch, $TACC_{90}$ stands for theoretically (i.e. calculated) precipitated calcium carbonate at a temperature of 90 °C. Values above 1.2 mmol/l significantly increase the risk of precipitation and scaling when heating drinking water. $TACC_{90}$ gives the first indication that the actual measurements have not been made.
	$PACC_k$	< 0.4 mmol/l	Translated from Dutch, $PACC_k$ stands for actual precipitated calcium carbonate observed in experimentally boiled and filtered water (cooled down) to eliminate carbon dioxide. When these measurements are possible, they offer a more complete picture than the calculated $TACC_{90}$, because they can also detect natural organic matter (NOM).
Other	TIC	> 2 mmol/l	To prevent a pH change in the network.

Source: Slaats *et al.* 2013.

Corrosion and soil aggressiveness

Water distribution underground infrastructure is also exposed to the external corrosion resulting from soil aggressiveness. This is specifically critical for metallic pipes,
which explains the need for external coatings and encasements (such as those illustrated in figures 5.27 and 5.28). The extent of this depends mostly on the soil composition and the moisture content. Wet and alkaline soils, such as clays, or swampy and
muddy ground with variable groundwater levels, will be more aggressive than those
in well-drained areas. Moreover, soils can be polluted by coalmine waste or various
other constituents that also contribute to the soil corrosiveness. The soil aggressiveness will be commonly assessed by measuring the following parameters:

- Soil resistivity (ohm-m),
- Soil pH (-),
- Oxidation-reduction (redox) potential (mV), and
- Concentrations of sulphides, chlorides, and sulphates (mg/kg).

Soil resistivity is the measure of its conductivity i.e. the capacity to resist the
flow of electricity, which is mostly influenced by the moisture and presence of different salts. More corrosive soils will have a lower resistivity, which is illustrated
in Table 6.12 (Robinson, 1993).

Table 6.12 Soil resistivity and corrosiveness

Soil resistivity (ohm-m)	Degree of corrosiveness
0 – 5	Severely corrosive
5 – 10	Very corrosive
10 – 30	Corrosive
30 – 100	Moderately corrosive
100 – 250	Slightly corrosive
> 250	Very slightly corrosive

Adapted from: Robinson, 1993.

Table 6.13 correlates the typical range of soil resistivity to the various types of
(contaminated) soils (Seidel and Lange, 2007).

The redox potential (ROP) characterizes the oxidation-reduction status of the
soil; it is essentially a measure of the soil aeration. Higher positive values of
ROP indicate aerobic conditions with higher oxygen levels, while negative values
reflect anaerobic conditions with greater soil microbial corrosion of metal pipes.

The evaluation of soil aggressiveness is commonly carried out using quantitative
methods by allocating weighting to each assessed parameter and arriving at a sum of
points that reflects the overall soil potential for pipe corrosion. This result is known
Soil corrosiveness in some literature as the *soil corrosiveness index* (SCI). A degree of subjectivity is
index to be taken into consideration in the absence of a generally accepted rating scale.
For example, Table 6.14 shows a broader list of soil parameters (Van der Zwan and
Blokland, 1989). By evaluating each factor as proposed in the table, the soil is:

- not aggressive if the sum of points is > 0,
- slightly aggressive if the sum is between -4 and 0,
- aggressive if the sum is between -10 and -5, and
- extremely aggressive if the sum is < -10.

Table 6.13 Soil resistivity of geological and waste materials

Material	Soil resistivity (ohm-m)	
	Minimum	Maximum
Gravel	50 (water saturated)	> 10^4 (dry)
Sand	50 (water saturated)	> 10^4 (dry)
Silt	20	50
Loam	30	100
Clay (wet)	5	30
Clay (dry)		> 10^3
Peat, humus, sludge	15	25
Sandstone	< 50 (wet, jointed)	> 10^5 (compact)
Limestone	100 (wet, jointed)	> 10^5 (compact)
Schist (rock of layers of various minerals)	50 (wet, jointed)	> 10^5 (compact)
Igneous and metamorphic rock	< 100 (weathered, wet)	> 10^6 (compact)
Rock salt	30 (wet)	> 10^6 (compact)
Domestic and industrial waste	< 1	> 10^3 (plastic)
Natural water	10	300
Seawater (35 ‰ NaCl)	0.25	
Saline water (brine)	< 0.15	

Source: Seidel and Lange, 2007.

Table 6.14 Assessment of soil aggressiveness

Parameters	Points	Parameters	Points
SOIL COMPOSITION		pH READING	
Chalky soil, sand	2	pH > 6	0
Chalky clay, marley chalk	1	pH ≤ 6	−2
Marley sand or clay, silty chalk	1	REDOX POTENTIAL (at pH=7)	
Silt, silty sand, clay, 75% mud	0	E > 400 mV	2
Peaty silt, Silty marl	−1	E = 200 - 400 mV	0
Clay, Marley clay, Humus	−2	E = 0 - 200 mV	−2
Marl, Thick silt	−3	E < 0 mV	−4
Muddy and swampy ground, Peat	−4		
SOIL CONDITION		CARBONATE CONCENTRATION	
Groundwater - none	0	> 5 %	2
Groundwater - permanent	−1	1 - 5 %	1
Groundwater - temporarily present	−2	< 1 %	0
Natural soil	0	H₂S AND SULPHIDE	
Backfill	−2	None	0
Same soil as trench	0	Traces	−2
Different soil to trench	−3	Present	−4
SOIL RESISTIVITY		COAL AND COKE	
> 100 ohm-m	0	None	0
50 - 100 ohm-m	−1	Present	−4
23 - 50 ohm-m	−2	CHLORIDES	
10 - 23 ohm-m	−3	< 100 mg/kg	0
< 10 ohm-m	−4	≥ 100 mg/kg	−1
WATER CONTENT		SULPHATES	
< 20 %	0	< 200 mg/kg	0
> 20 %	1	200 - 500 mg/kg	−1
		500 - 1000 mg/kg	−2
		> 1000 mg/kg	−3

Adapted from: Van der Zwan and Blokland, 1989.

Table 6.15 Soil evaluation for DI pipe encasement (DIPRA, 2017)

Parameters	Points	Parameters	Points
SOIL RESISTIVITY*		REDOX POTENTIAL	
< 15 ohm-m	10	E > 100 mV	0
≥ 15 – 18 ohm-m	8	E = 50 - 100 mV	3.5
> 18 – 21 ohm-m	5	E = 0 - 50 mV	4
> 21 – 25 ohm-m	2	E < 0 mV	5
> 25 – 30 ohm-m	1	SULPHIDES	
> 30 ohm-m	0	Positive	3.5
pH READING		Traces	2
0 – 2	5	Negative	0
≥ 2 – 4	3	MOISTURE	
> 4 – 6.5	0	Poor drainage, continuously wet	2
> 6.5 – 7.5**	0	Fair drainage, generally moist	1
> 7.5 – 8.5	0	Good drainage, generally dry	-2
> 8.5	3		

* - based on the water-saturated box method.
** - for sulphides present, and E < 100 mV or negative, 3 points should be given in this range.

The Ductile Iron Pipe Research Association (DIPRA, 2017) suggests a 10-point scoring method (based on the AWWA-C105 standard) to assess the need for polyethylene sleeves around the pipe, according to the points given in Table 6.15. The soil is considered aggressive when it has a sum of points of 10 or more, with potential to corrode DI pipes.

Corrosion control

The ways of achieving corrosion control are (Brandon, 1984):

1) selection of adequate materials and design concepts,
2) modification of water quality parameters,
3) use of inhibitors,
4) cathodic protection,
5) use of corrosion-resistant linings, coatings and paints.

Material selection - A careful material choice can reduce corrosion, though it is just one of the aspects to be taken into account. Several alternative materials are usually compared based on the cost, availability, ease of installation and maintenance, as well as resistance to corrosion. Compatible materials should be used throughout the system as much as possible. Where this is difficult, galvanic corrosion can be avoided by placing insulating couplings made of different materials between the pipes.

Network design - Poor design of the pipes and structures may cause severe corrosion even in materials that may be highly resistant. Some of the important design considerations include:

• avoiding dead ends and stagnant areas,
• provision of adequate drainage where needed,

- selection of an appropriate flow velocity,
- selection of an appropriate metal thickness,
- reduction of mechanical stresses,
- avoiding uneven heat distribution,
- avoiding sharp bends and elbows,
- provision of adequate insulation,
- the elimination of grounding electrical circuits in the system, and
- providing easy access to the structure for periodic inspection, maintenance and replacement of damaged parts.

Water quality adjustment -This is the easiest and most practical way to make water non-corrosive. However, it is not always effective bearing in mind possible differences in water quality at the sources. Two basic methods are: pH correction and oxygen reduction.

Most of the corrosion occurs at low pH values. Chemicals commonly used for pH adjustment are:

1) lime, as $Ca(OH)_2$,
2) caustic soda, NaOH,
3) soda ash, Na_2CO_3, and
4) sodium bicarbonate, $NaHCO_3$.

Oxygen is an important corrosive agent because it can act as an electron acceptor; it reacts with hydrogen and also with iron ions. These are all processes that promote corrosion, thus the reduction of oxygen can lessen their effects. Oxygen removal is relatively expensive but some control measures can be introduced through the optimization of aeration processes and the sizing of groundwater wells and pumps to avoid air entrainment.

Inhibitors - Chemicals added to the water that form a protective film on the pipe surface are called inhibitors. This film provides a barrier between the water and the pipe, which reduces corrosion. Various products are used for this purpose, which can be classified into three main groups: *Inhibitors*

1) chemicals which cause $CaCO_3$ formation,
2) inorganic phosphates, and
3) sodium silicate.

The difficulties in using inhibitors lies in the control of the process. An interrupted supply of the chemical can cause dissolution of the film and too low a dosage results in a fragmentary film, both of which increase pitting. On the other hand, excessive use of some alkaline inhibitors can cause an undesirable build-up of scale, particularly in harder waters. Finally, the flow rates must be sufficient to transport the inhibitor to all parts of the pipe surface; otherwise an effective film will not be formed.

Cathodic protection - Cathodic protection is an electrical method for preventing metallic corrosion. It forces the protected metal to behave as a cathode and *Cathodic protection* therefore unable to release electrons. Basic methods of applying cathodic protection are:

1) the use of inert electrodes (with high level of silicon cast iron or graphite) powered by an external source which forces them to act as anodes,
2) the use of magnesium or zinc as anodes that produces a galvanic reaction with the pipe material. Being more reactive than iron, they corrode, thereby keeping the pipe protected (*sacrificial corrosion*).

Sacrificial corrosion

Cathodic protection is expensive; it has to be controlled and renewed after a period of time. In addition it has little application in localities where corrosion has already started (holes, crevices, etc.). Two examples of its application on steel pipes passing under water streams are shown in Figure 6.59.

Figure 6.59 Cathodic protection applied to steel pipes

Linings, coatings and paints - Corrosion can be kept away from the pipe wall if it is lined with a protective coating. The linings are usually applied mechanically, either during the manufacturing process or before pipe laying. They can also be applied to pipes in service, which is much more expensive. The most common protective materials are:

1) epoxy paints,
2) cement mortar, and
3) polyethylene.

Epoxy paints are used for steel and DI pipes. These are smooth coatings that have no effect on the water quality, but are relatively expensive and less resistant to abrasion. Cement mortar is a standard lining for DI pipes, and also used for steel and cast iron pipes. It is relatively cheap and easy in application. The disadvantage is the relatively thick coating required which reduces the carrying capacity of the pipe. The rigidity of the lining may also lead to cracking or sloughing. Finally, a polyethylene coating is used in DI and steel pipes. It is a durable material with excellent characteristics (smooth and resistant) but also relatively expensive.

Some inconsistency exists in the terminology; the term lining, when considered as a process, refers to the internal pipe protection, while coating refers to its external protection. However, lining can also be understood as a separate internal layer inserted into the pipe (usually of a different material to the pipe itself) and attached mechanically to the pipe wall. This process is then also known as *structural (re)lining*. In the same context, the term coating (internal or external) refers to a thinner layer of protection applied by spraying or by brush, which adheres to the pipe material directly and becomes an integral part of it i.e. cannot be removed easily.

Various linings are more commonly used as pipe repair techniques, analysed versus the replacement option; this is further discussed in Section 6.2.4. Examples of application of various linings/coatings are shown in figures 4.41 to 4.43 for steel pipes. Two examples of DI pipes are shown in Figure 6.60.

Figure 6.60 Cement-lined DI pipes with (left) epoxy and (right) PE external coatings

6.2   Network maintenance

Network operation and maintenance are often interrelated. Effective operating systems that meet the consumer's requirements are also likely to reduce the level of maintenance. The maintenance considered in this context is predominantly a response to e.g. pipe bursts, and is referred to as *reactive maintenance*.

Reactive and preventive maintenance

On the other hand, proper maintenance also contributes to the optimal operation of the system. As such, it is more of a condition, or requirement, for effective operation, and is therefore known as *preventive maintenance*.

Just as with the operation, efficient maintenance relies largely on effective monitoring of the system. This is illustrated by the following example (Vreeburg and Van den Hoven, 1994). Figures 6.61 and 6.62 show the monitoring of turbidity in one CI pipe. In the first case, a higher turbidity is registered

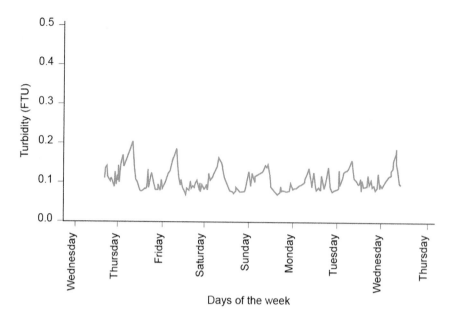

Figure 6.61 Turbidity during night flow

Source: Vreeburg and Van den Hoven, 1994.

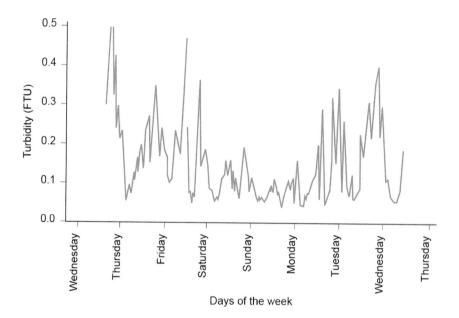

Figure 6.62 Turbidity during day flow

Source: Vreeburg and Van den Hoven, 1994.

overnight, during longer retention times (low consumption), which indicates the release of corrosion products as a source of the problem. A possible remedy in this case is the control of the retention times through modification of the operation. When this problem occurs on a large scale, relining or pipe replacement should be carried out.

The next figure shows the turbidity in larger amplitudes, and higher during the day. The reason here is entirely different: high velocity (high consumption), which is causing re-suspension of the sediments in the pipe. The maintenance action taken in this case should be flushing of the pipe and, if necessary, coating renewal.

6.2.1   Planning of maintenance

The selection of the type and level of maintenance follows an overall asset management strategy based on two main principles:

- standard of service to the consumer should be regarded as the primary objective,
- within the constraints set by the standards of service, decisions should take into consideration the economic lifetime of the network components.

The problems caused by the network deterioration determine the form of the strategy. Their thorough description and complete understanding of them is a prerequisite. The strategy should contain all the steps necessary for proper economic decisions. It should be future-oriented and therefore not only able to cope with current problems. It should also be sufficiently flexible to allow for easy incorporation of improvements in technology. The major stages of the strategy to be followed are listed in Figure 6.63 (Brandon, 1984).

Practical factors that influence the strategy have already been discussed in previous sections and here are only summarised:

1) design and technical layout,
2) soil conditions,
3) surface activities,
4) climate,
5) water quality,
6) material selection,
7) construction methods, and
8) operational pressures and flows.

A preference for either preventive or reactive maintenance is derived from the strategy selected. Generally speaking, the annual costs of repairs and cleaning operations spent responding to consumer complaints are lower than the annual cost of the investment necessary for main rehabilitation and replacement. However, this is only true for the standard frequency of pipe bursts. The expected trend is that the future number of ruptures will increase over time. Preventive maintenance can extend the economic lifetime of the system, and therefore is a 'must'; only the level of maintenance is debatable.

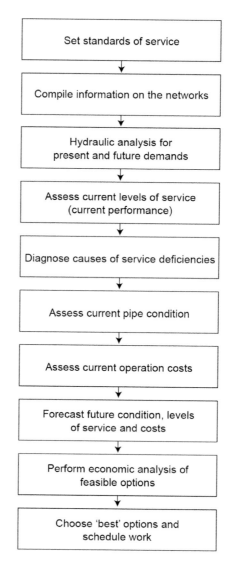

Figure 6.63. Organisation of maintenance

Source: Brandon, 1984.

6.2.2 Pipe cleaning

When disturbed, corrosion deposits in pipes or sediments caused by improperly executed treatment have to be removed in order to prevent water quality deterioration. The decision to conduct pipe cleaning is a result of the following situations:

- consumer complaints about water quality (colour, turbidity),
- after a new pipe has been laid or an existing pipe has been repaired,

- the need for removal of excessive disinfectant used to kill bacteria or living organisms in pipes, and
- systematic cleaning as a part of regular (preventive) network maintenance.

Three techniques for pipe cleaning are commonly used: flushing, air scouring, and swabbing (or pigging).

Flushing

By opening a hydrant or washout on the main, an increased water flow is generated to remove loose deposits (an example of application is shown in Figure 6.64).

Figure 6.64. Pipe flushing process.

Based on the type of deposits in the pipes, a minimum velocity needs to be generated for a certain period of time in order to remove them. Various guidelines exist in this respect. Recommendations in earlier practice in Great Britain are shown in Table 6.16 for the transport of sand particles of average size d = 0.2 mm (Brandon, 1984). These suggest extreme amounts of water for the cleaning process. Flushing for instance a pipe of 200 mm in diameter for only one minute would demand nearly 5 m³ of water.

Table 6.16 Required flushing velocity for sand particles, d = 0.2 mm

Pipe diameter (mm)	v (m/s)	Q (l/s)
50	1.3	2.7
75	1.6	7.2
100	1.8	15.0
150	2.2	41.0
200	2.6	83.0

Source: Brandon, 1984.

More recent guidelines for flushing velocities applied in the Netherlands are generally lower: around 1.5 m/s, due to finer sediment and the use of mostly corrosion-free pipes in the system. As a general guideline, the approximate quantity of water needed is equivalent to three full volumes of the pipe that is being flushed. Taking the same diameter of 200 mm on a short section of, say, 100 m would require a flushing volume of $(0.2^2 \times 3.14/4) \times 100 \times 3 = 9.42$ m^3, which should be flushed by a velocity of 1.5 m/s (≈ 47 l/s) in 200 seconds (= 300 m divided by 1.5 m/s).

The example in Figure 6.65 shows monitoring of the effects of flushing in a pipe section which was conducted in the area of Rotterdam. The diagram shows the recorded pressures, velocities and turbidity, and indicates the second and third phase of flushing, each lasting approximately 20-25 minutes with a break of a few minutes in between. As a result of the hydrant opening, the normal service pressure of approximately 28 mwc (= 2.8 bar) drops to 10 mwc (= 1 bar) during flushing, while the flushing velocity will reach approximately 1.3 m/s. In this particular section, the initial turbidity of 64 ftu generated in the first phase was brought down to approximately 25 ftu in the second phase and eventually to approximately 15 ftu in the third phase. The pressure surge of 40-45 mwc is a consequence of closing the hydrant, which should be conducted slowly.

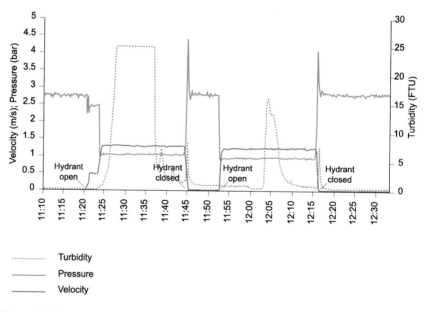

Figure 6.65 Example of a pipe flushing diagram

Courtesy: Europoort, Rotterdam, 1997.

The US standards suggest even lower flushing velocities that are to be maintained above 0.8 m/s and preferably not below 1.1 m/s, with two to three volumes being flushed. The corresponding guidelines are given in Table 6.17 (AWWA, 2016-1).

Table 6.17 Flushing guidelines at v ≈ 0.8 m/s and p/ρg = 28 mwc

Pipe diameter (mm)	Q (l/s)	No hydrants (65 mm)
100	6.3	1
150	12.6	1
200	25.2	1
250	37.9	1
300	56.8	2
400	100.9	2

Adapted from: AWWA, 2016-1.

Under the indicated pressure of 28 mwc (40 psi), one standard hydrant with a nozzle of 65 mm (2.5 inch) discharges approximately 60 l/s (1000 gpm). Hence, for flushing of larger diameters, two hydrants of this type need to be opened in series.

Flushing is a simple method of cleaning, but not always efficient. The disadvantages are:

- large amounts of water used (particularly in large diameters),
- large number of hydrants opened at the same time may cause a significant drop of pressure with potential for backflow contamination,
- the velocity increase in the pipe being flushed may disturb the flow and pressure pattern upstream of the cleaned section,
- in areas with progressive corrosion, flushing may offer only a partial improvement, and
- not all parts of the distribution system may be equally suitable for the generation of high velocities (e.g. in low pressure areas).

Flushing pipes as a preventative measure requires careful planning. The following factors need to be taken into consideration:

- selection of the optimal pipe route and the sequence of flushing,
- the location of valves that are operated in order to isolate the flushed route from the rest of the system,
- total length of section that is flushed in one run, and
- choice of hydrants (number and location) that will have to be opened simultaneously in order to generate the necessary velocity.

Taking into consideration that the process is water-intensive, the choice can be made only to flush the more critical parts (more frequently), rather than the whole system periodically. The decision in this case will be based on regular monitoring of turbidity in the network. The target of any flushing plan in any scenario is to clean the system efficiently i.e. with the minimum quantity of water possible, as well as with the minimal operation of hydrants and valves. However, particularly in looped systems, making an optimal flushing plan is almost impossible without the use of a well-calibrated network computer model. The low-velocity areas pinpointed in the simulation results also need to be more carefully assessed for turbidity, leading to more frequent cleaning.

The execution of the flushing programme requires strict safety measures and the customers should be warned against possible discoloration of the water (if doing laundry, for instance). The level of disturbance on the street will be lower if the procedure takes place overnight when it may also be easier to generate the required flushing velocities.

Air scouring

In situations where water quantities available for pipe cleaning are limited, air scouring can be used as an alternative method to flushing. In this method, compressed air is injected into a continuous flow of water. Pushed by the air, the water will form into discrete slugs forced along the pipe at high velocities. The method is illustrated in Figure 6.66.

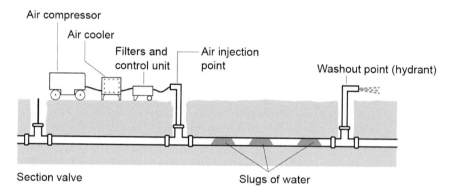

Figure 6.66 Air-scouring equipment and procedure

The atmospheric air injected at intervals of a couple of seconds is filtered and cooled prior to the injection. A common injection point is at the hydrant but a service connection can also be used for this purpose. The length of the section that can effectively be cleaned is normally a few hundred meters.

The discharged water is left to spill around and will be collected by the sewer system. The jet of water leaving the hydrant is usually quite strong; it is therefore important that the hose installed on the hydrant is securely fixed and ends with a protective plastic sheet or a box that dampens the vibrations (Figure 6.67).

Compared to ordinary flushing, air scouring requires both more equipment and energy supply. Control of the process may be a problem in some cases; the decision on optimal injection frequency, duration and the pressure is often empirical.

Figure 6.67 Discharge point with dampening of the water jet

Swabbing

Swabbing is a cleaning technique by which the deposit is mechanically removed from the pipe. A cylindrical swab is inserted into a pipe and driven along by the water pressure thereby pushing the deposits ahead. The swab is porous and allows some water flow to pass, which enables the transport of deposits.

Swabs are usually made of polyurethane of different hardness. The hard grade is normally used for smaller reductions in pipe cross sections (up to 30 %). The soft grade is more convenient for larger reductions as it is compressible and can contract when it meets an obstruction in the main. When it is not possible to remove the obstruction, a soft swab is likely to be torn into small pieces, not further clogging the pipe.

Swabs made of abrasive materials are called pigs. They are stronger in construction and can be used for more aggressive cleaning (pigging). However, this should be avoided in the case of severe tuberculation as weak sections of the pipe could crack.

Different models of swabs and pigs are shown in Figure 6.68.

Figure 6.68 Types of swabs and pigs

Manufacturer: Kleiss & Co.

Swabs are normally oversized, compared to the corresponding pipe diameter (up to 25 % larger). The process will be repeated with a number of units, and a visual inspection of the water at the end of the section, as well as the condition of the swab, will indicate the success of the cleaning process.

Pipe diameters up to ± 1000 mm can be treated with this method. Fire hydrants are the most common entry and exit points for small swabs while special fittings should be installed in the case of large diameters. The velocity of the swab will be similar to the normal operating velocities in the pipes.

Swabbing can be an advantageous method of pipe cleaning in water-scarce areas, compared to flushing or air scouring. Thorough preparations and monitoring of the process are needed in this alternative. Apart from being destroyed, swabs can be lost or jammed in the case of branches or connections that draw more water than the section that is being cleaned. More expensive models contain a sensor so that their location in the pipe can be detected at any time. Nonetheless, proper care should be taken to isolate the cleaning section. A reversible flow may help in releasing a jammed swab.

6.2.3   Animal disinfection

Water quality in distribution pipes and reservoirs can be affected by the appearance of various types of organisms. Based on their preferred habitat, these can be classified into three groups:

1)   those which swim freely; examples are Gammarus and Cyclops,
2)   those that live in deposits, such as Chiranomid larvae,
3)   those that are attached to the pipe surface; an example is Asellus, an organism that feeds on non-living organic matter.

The density and composition of animal populations in water distribution systems vary widely, and usually result from improper water treatment, insufficient disinfection, poor conditions of the pipes, insufficient hygienic precautions during the repair or replacement process, or poor maintenance and lack of protection of the service reservoirs. For instance, Assellus typically occurs in the case of ineffective treatment of raw water containing algae, whilst Chiranomid larvae usually occur as a result of flying insects entering badly protected openings on the reservoirs.

The problem is predominantly aesthetic and it is therefore a matter of maintaining animal numbers at levels 'invisible' to consumers. In some cases an aquatic organism can act as a host for parasites such as Cyclops, which can transmit the guinea worm, a dangerous parasite that appears in tropical and subtropical countries. However, the risk of infection is low in piped systems that convey treated water. As most of the indications come from complaints, the assessment of animal presence through sampling is advisable as a preventive measure. This will be done by flushing water thorough a hydrant at a controlled flow rate through fine-meshed sieves and estimating the numbers by an order of magnitude. Examples of sampling sieves are shown in Figure 6.69.

De-infestation can be done either by cleaning or by chemical treatment. Swimming and settled animals are relatively easily flushed out while those attached to the pipe have to be dislodged first, before being flushed. This can be done by swabbing or using the air scouring technique.

Chemical treatment is carried out where flushing is insufficient. The chemicals commonly employed are chlorine and occasionally pyrethrins and permethrin. When using chlorine, higher concentrations are required. Maintaining 0.5-1.0 mg/l of residual chlorine for a week or two will be sufficient in most cases (Brandon, 1984). In isolated and extreme cases, much higher dosages of 10-50 mg/l can be used, specifically for cleaning the service reservoirs but these should be disconnected from operation. Concentrations applied during pre-chlorination may be effective in reducing the appearance of animals in the treatment works.

Pyrethrins and permethrin are types of pesticides that can be used effectively against Asellus, Gammarus and Chiranomid larvae. The recommended dose is 10 µg/l, which can be increased up to 20 µg/l, if necessary. The treatment should be carried out with the utmost caution and in parts of the system fully isolated by valves. The normal contact time is 24 hours after which the pipes should be flushed with clean water; the recommended volume is twice the volume of the pipe (WHO, 2004). The flushed water should not be disposed of into natural streams because both substances are toxic to fish.

Figure 6.69 Sampling sieves for animal detection

Courtesy: Waternet.

Long-term measures include the removal of organic matters (by restricting food to the animals), which can be achieved by the following methods (Brandon, 1984):

1) improvement of the treatment process regarding suspended solids removal and animal penetration,
2) periodic cleaning of pipes and service reservoirs,
3) maintenance of a chlorine residual throughout the distribution system,
4) proper protection of openings on service reservoirs (i.e. ventilators with nets and manhole covers), and
5) the elimination of dead ends and stagnant waters wherever possible.

In addition, this problem is another reason why the distribution network needs be kept under sufficient and continuous pressure in order to prevent the ingress of contamination. Finally, all maintenance and repair works should be conducted respecting strict procedures and hygienic codes, which includes both the distribution network and the treatment works (in the latter, for example, the procedure for backwashing rapid sand filters).

6.2.4   Pipe repairs

Pipe repairs can be classified into two groups:

1) for damage caused by pipe transportation and handling at the site, and
2) for damage in service.

The first group includes correction of the pipe cross-section deformation (pipe re-rounding) and repairs to external and internal coatings; examples are shown in figures 6.70 and 6.71. An example of a typical crack in the internal cement lining of a steel pipe is shown in Figure 6.72.

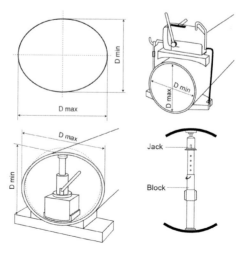

Figure 6.70 Correction of pipe deformation
Source: Pont-a-Mousson, 1992.

Figure 6.71 Repair of internal cement lining

Figure 6.72 Repair of the internal cement lining of a steel pipe

Repairs to pipes whilst in service can be executed at the locality (for circumferential failures and small holes) or along a section (for longitudinal splits and blowouts). Along a section is more complicated and sometimes requires replacement of the whole pipe. The use of *trench-less technology* may be considered here. By using appropriate equipment, the damaged pipe will be simply broken into pieces and the rubble pulled out, leaving a tunnel for the new pipe. Owing to their strength and flexibility, PE pipes are often chosen in this situation.

Open-cut replacement follows an identical procedure to the laying of new pipes. This is a more expensive alternative but offers a more flexible choice of material and size of pipe. An example of an open-cut section replacement is presented in Figure 6.73. The new piece of pipe is cut at a shorter length (L<C) to allow the minimum space (J) for the coupling. A few examples of couplings are shown in Figure 6.74.

Less damaged pipes can be repaired by relining. Conventional cement and epoxy are coatings exclusively used to prevent corrosion and have no structural function in the pipe. Cement relining is cheaper but less uniform in terms of results. The consequence is that the work on some sections may have to be repeated after the inspection has been carried out while at others, clogging of service connections might occur. An example of such equipment which can be applied to pipes above 500 mm in diameter is shown in Figure 6.75.

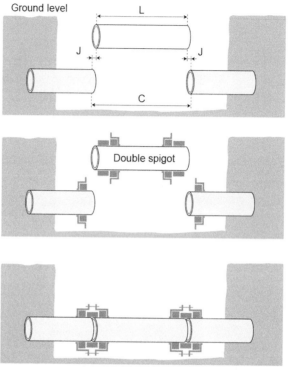

Figure 6.73 Pipe repair - section replacement

Figure 6.74 Couplings used to connect pipes of various materials and diameters

Figure 6.75 Equipment for cement relining of pipes

Manufacturer: Aegis.

Really structural linings involve the installation of various types of hoses or pipes within the old pipe that will be tightly fitted to its wall. Such a solution greatly improves the strength of the pipe. Reduction in the original diameter, caused by the placing of an inner pipe, can be offset by the improved roughness so that this measure does not necessarily result in flow reduction (Figure 6.76).

Figure 6.76 Structural relining of an old CI pipe

The decision on the repair action depends on an overall assessment of the pipe condition for which visual inspection is the most suitable method. This is of course not easy because the pipes are buried in the ground. Small cameras on wheels are commonly used that are operated from the surface thorough an emptied pipe. An example of this type of equipment is shown in figures 6.77 and 6.78.

Figure 6.77 Pipe inspection camera

Figure 6.78 Ultrasonic equipment for inspecting the cement lining of reinforced concrete pipes

Interventions in the network are often conducted without interruptions to the service. This includes the quick repair of smaller pipe bursts, the installation of a service connection, and in more extreme cases the insertion of valves in the pipes under pressure. Examples of these appurtenances are shown in Figure 6.79.

Figure 6.79 Parts for quick repairs

6.3   Organisation of the water company

The way in which operation and maintenance tasks are implemented has implications for the organisational set-up of the water company. The availability of technical means and expertise in itself will not be sufficient if appropriate planning of activities and coordination between various services are missing.

The entire management of the water distribution is usually taken care of by a department that is part of a larger water supply company[4]. Such a department has to function in line with the general policies of the company, providing well-differentiated tasks and responsibilities among the employees.

6.3.1   Tasks

The activities of the distribution department (company) are basically divided into office work and fieldwork, some of these being centralised. Hence, the global structure consists of the head office and a number of district centres. The most important tasks of the central office are:

*   collection of technical data (mapping),
*   design and computer modelling of the main network,

---

4 In a larger set-up, water companies can also include sewerage, wastewater collection and treatment; thus, covering the entire urban water cycle in which case the name 'water utility' covers the wider scope of activities more accurately than the name 'water supply company'.

- financial aspects of the network design and operation,
- monitoring of the network operation (control centre),
- control of major consumers,
- remote meter reading and billing of consumers, and
- central administration and revenue collection.

More practical responsibilities of the district centres are:

- construction of the network and service connections,
- preventive maintenance (repair and cleaning),
- failure service,
- installation and maintenance of water meters,
- leakage detection and repair,
- water quality sampling control,
- control of indoor installations,
- connection and disconnection of the consumers,
- management of the stock of spare parts,
- measurements in the network,
- registration of technical data, and
- administration of the activities.

Depending on the area supplied and specific procedures, some tasks can be real-located between the district centre and the head office. In the case of smaller companies, the major infrastructural works will be outsourced to external con-tractors, and also the water quality control can be centrally supervised for a few water companies, in order to provide sufficient benchmarking. Lastly, sev-eral water supply companies can collaborate in revenue collection providing integral billing of their services (electricity, gas, water, etc. in one bill for the consumers).

6.3.2 Mapping

Maps are the starting point of any maintenance. A detailed and regularly updated database is therefore of paramount importance. The scale and level of information will be dictated by the purpose of the map.

Maps used for design by means of computer modelling consist of distribution pipes usually with a diameter larger than 80 mm. The main information offered here concerns pipe routes, diameters and materials. Pipe junctions and crossings, as well as the position of other components (in particular, valves) have to be clearly indicated. A common map scale for this purpose is 1:1000 to 1:5000. Such maps will be digitised for computer use, providing quick access and easy correc-tion and storage. An example is shown in Figure 6.80.

Maps for maintenance purposes are on a smaller scale and have more detailed information. These have to show the location of service pipes, valves, domes-tic connections, etc. enabling an efficient response when problems occur. If made by computer, the maps can be created with different sorts of information and overlaid, which is convenient for analysis (e.g. separate maps showing the

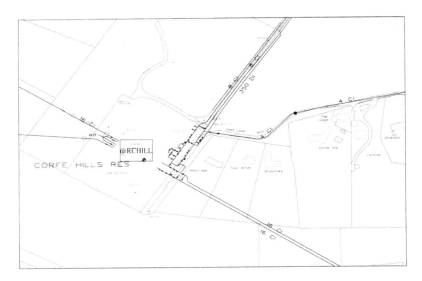

Figure 6.80 Computer-drawn map of a network section

Source: Wessex Water, 1993.

topography, houses and streets, pipe network, sewer network, electricity and gas, etc.). Not that long ago, maps would be commonly drawn on transparent paper (as can be seen in the example shown in Figure 6.81). Unfortunately, hand-drawn maps are still the primary source of information on the distribution network in many water companies in developing countries, which means they may be relying on incomplete or incorrect information based on the empirical knowledge of local operators.

The recent introduction of geographical information systems (GIS) into water distribution system management has greatly enhanced the amount and accuracy of data available. GIS maps have become readily available and can receive any additional information that becomes available after any replacement, connection or disconnection, or expansion of the system has taken place. In this way, these maps enable multiple use: providing direct input for the computer model, accurate billing information, and the location of system components that are malfunctioning and have to be repaired, etc. An example is shown in Figure 6.82.

Typical categories of GIS information available in databases used by water supply companies are shown in Figure 6.83; the map layers per category are shown in Table 6.18 (AWWA-2, 2016).

Data collection for GIS databases can be tedious work. In every case, every new component should be registered with the coordinates at the moment of laying, as is the case in the example shown in Figure 6.84. This also includes any repair or replacement works.

The retroactive efforts include close inspection of all available information, with frequent field visits where the location and condition of the components

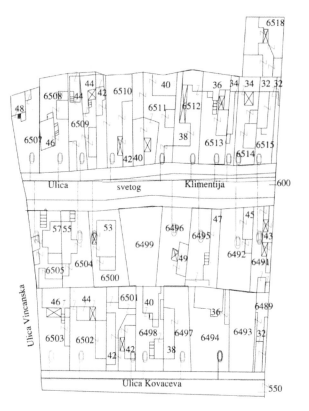

Figure 6.81 A hand-drawn map of a residential area

Source: Obradović, 1991.

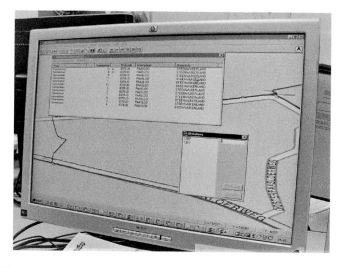

Figure 6.82 A modern network information system for scheduling maintenance activities

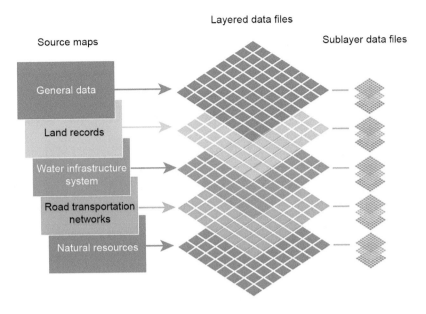

Figure 6.83 Layers of database information categories

Adapted from: AWWA, 2016-2.

Table 6.18 Typical data sets used by water supply companies

Data category	Example map layers
General data	Control information Planimetric features Hydrology features
Land records	Property boundaries Easements Right-of-ways
Natural resources	Groundwater data Drainage data Soil data Flood plain boundaries Topographical features Vegetation information
Water infrastructure system	Water piping Water valves and utility holes Service areas Main facilities (wells, pump stations, tanks, treatment plants) Other utilities
Road transportation networks	Road centre lines Pavement locations Road intersections Bridges

Adapted from: AWWA, 2016-2.

Figure 6.84 Registration of the network component for the GIS database

should be verified using pipe detectors and cameras. Visual inspection during regular maintenance, or reactive maintenance when the excavations need to take place anyway, is also a useful opportunity to learn more about the system.

The completed database can then be used for the extraction and plotting of various information. An example in Figure 6.85 shows the GIS map depicting various pipe ages in the distribution network of Belgrade, ranging up to 100-year-old pipes. This is then used for various purposes, such as the pipe roughness estimate in the network computer modelling, projection of the asset value, the remaining economic lifetime, and the optimal timing for the pipe replacement, as well as the selection of an adequate pipe cleaning programme. A similar map showing water metering of service connections is the primary source for billing of consumers but is also effectively used in the design of nodal demands and calibration of computer models.

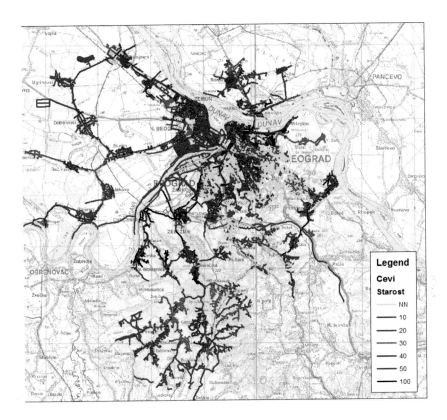

Figure 6.85 A GIS map showing different pipe ages in years, in the distribution network of Belgrade

Source: BWS, 2013.

6.3.3 Structure and size

Construction

Construction work in the system can be carried out by the company or by a hired contractor. This decision determines the degree of investment required in manpower and mechanisation.

Having work carried out by specialised companies has the following advantages:

- engaged labour and mechanisation is always appropriate to the amount of work,
- the work of the contractor can be partly carried out in conjunction with other services (gas, electricity, etc.) which reduces the costs to the water distribution company.

Despite the fact that labour costs in the developing world are relatively cheap, keeping an in-house construction department would appear to be fairly expensive, due to the inefficient employment of resources. This is an option only in the case of large utilities.

Maintenance

This is usually the largest department in the company. The number of employees is related to the size of the network but also to the required level of maintenance; loss of labour due to illness or holiday should also be taken into account. Contractors can sometimes be appointed to implement particular maintenance tasks which are not regular and may require specific equipment (examples shown in figures 6.75 and 6.78).

Failure service

The number of people required is dependent on the intended goals to be achieved. Where a 24-hour supply is an objective, shifts are necessary. Mobile equipment and quick access to all parts of the network are required and distances, topography, quality of the roads, etc. have to be taken into account.

Dutch water supply companies use well equipped vehicles for quick interventions, as the one shown in Figure 6.86. The van is used 24/7 and the worker on duty must stay within a certain area if the shift is outside regular working hours; the vehicle is then parked at his home address. They also have a laptop or a tablet at their disposal for quick access to the needed information in the case of an alarm that can be received at any moment. During normal working hours regular maintenance is scheduled for when there are no calamities.

Figure 6.86 Mobile workshop for a quick repair service

Investment in equipment, as well as proper planning of the activity, can reduce the number of personnel required. This also includes a stock of spare parts for timely execution of necessary works in the network. One example is shown in Figure 6.87.

Figure 6.87 Centralised stock of spare parts

Courtesy: Waternet.

Supervision

The quality of workmanship has to be permanently supervised. A small number of qualified staff is involved here, responsible also for formulation of conditions, such as:

- selection of the materials,
- standardisation of working procedures,
- standardisation of control procedures (testing of installations, conditions for connecting and disconnecting, etc.).

Supervision can sometimes be carried out on location in the district by the head office.

Administration

This is a supporting element of the company. An accurate, not necessarily extensive, record on the management of the company must be kept. The number and profile of the people depends on the type of the activity. Where there is a head office with administrative staff, most of these tasks are carried out there. However, at least the technical records of all the connections are kept in the district centre (for the failure service), together with administrative records on the employees, use of the materials, equipment, etc.

6.3.4   Example

An example of typical water distribution management in the Netherlands is illustrated in this section. Water Supply Company 'Drenthe' (WMD) was founded in 1937 with a small distribution system serving a few thousand people from less than one hundred connections in a rural area. Nowadays it is the biggest water company in the province. Basic information about the company is given in Table 6.19 (statistics in 2001).

Table 6.19 Water Supply Company 'Drenthe' – general data (2001)

Area covered	2200	km²
Population served	450,000	
Number of connections	180,000	
Distribution of pipelines	4000	km
Service pipes	4800	km
Production	32,000,000	m³
Average day	90,000	m³
Maximum day	152,000	m³
Number of employees	179	
Predominate pipe materials	PVC-54 % AC-38 %	

Organization setup

The company is non-profit based, and owned by the municipalities supplied and the province (50-50). Representatives of each group, together with the company director, form the Board of Supervisors, which is the actual governing body. The structure of the company is shown in Table 6.20.

Table 6.20 Water Supply Company 'Drenthe' - structure of the company (2001)

Departments	Head Office	Districts	Total
Director	1	0	1
Production	19	8-workshop/18-PST	45
Distribution	2	94	96
Finance	13	0	13
Personnel	24	0	24
Total	59	120	179

   The organisation, headed by a director, consists of four departments: production, distribution, finance and personnel. The department heads report to the director. Over 50 % of employees work in the distribution department, which is divided into three districts. One of the district centres looks as described in Table 6.21.

Table 6.21 Water Supply Company 'Drenthe' – sample district

Area covered	700	km²
Population served	150,000	
Number of connections	60,000	
Distribution pipelines	1300	km
Production in 2001	12,000,000	m³
Number of staff	27	
Manager/deputy manager	1/1	
Supervisors	3	
Inspector	1	
Fitters	18	
Storekeeper	1	
Administrator	2	

Annual investment in the district network is in the order of EUR 560,000 with approximately 800 new connections and 15 km of pipeline. The value of the stock is approximately EUR 80,000.

Activities in the design phase

Each project, after getting approval from the Board of Directors, goes to the planning stage. This mainly comprises the following data collection and necessary investigations:

Socio-economic survey - The data are available in the municipality and other corresponding institutions.

Topographic survey - Supplied from other sources.

Soil investigation - No regular soil investigation is carried out; only special cases can be considered (e.g. road crossings).

Existing services - Other services in the area are closely coordinated. In the case of new extensions, it is common practice to lay all the service lines in the same trench and at the same time. Maps with precise routes for all the services are available for any interested organisation or individual.

Design - In most cases the design is carried out within the distribution section of the company by experienced technical staff with sufficient working facilities (a senior design engineer and a number of assistants). Approximately 80 km of pipelines is designed per year, on average. All the data and originals of the computerised drawings are kept in the head office. The district offices can access the database at any time. The distribution section services the connection designs as well.

Bearing in mind the soil conditions, PVC is the most preferred pipe material for up to 500 mm diameter. The minimum design pressure is 25 mwc above street level. Leakage of 10 % is taken into account. The demand variations are represented by a peak factor of 1.50 for the maximum consumption day, and 1.68 for the maximum hour consumption. Other common design standards are:

- domestic connection, D = 25 mm,
- < 20 connections, D = 80 mm,
- > 20 connections, D = 100 mm,
- design velocities, v = 0.5-1.5 m/s, and
- fire hydrants, each 100-200 m.

Construction works

In 1955 the company covered approximately 26,000 connections and employed 120 workers (fitters and diggers), plus approximately 200 on a temporary basis. Nowadays this number is drastically reduced; smaller construction works are carried out within the company by the maintenance section while for larger activities a contractor is engaged.

Stock of spare parts - Each district has its own storage for pipes, fittings and other equipment and materials. Transportation to the location is carried out by the contractor.

Quality control - All the components in the system carry the KIWA certificate. Any damage to the pipes caused during the manufacturing process is repaired at the manufacturer's cost, no matter when the pipe was produced.

Pipe laying - The soil is usually sandy, therefore it is not usual to place extra bedding under the pipe. PVC distribution pipes are jointed by dual socket 'push-in' joints. Each pipe is cleaned before laying. The backfilling is compacted by mechanical vibrators. The pipeline is usually tested after the backfilling, at a pressure of 90 mwc, for two hours. Pipelines are disinfected with chlorine before commissioning.

Work supervision - One field engineer is responsible for the full-time supervision of the pipe laying. They are provided with a car and wireless communication system. The contractor is totally responsible for the quality of the work. Supervision of smaller jobs is carried out part-time.

Service connections - In most cases the connections are made by the company's fitters but for new residential areas with many service lines this is carried out by a contractor. All the components of the service connections are supplied from company stock (a KIWA certificate is required here as well). One field technician is responsible for the work supervision.

Operation and maintenance

As-built drawings - This information is computer-processed. The detailed position and description of the valves, fittings, washouts, pits, crossings, etc. is indicated on a map layout, with all the information necessary for maintenance activities.

Pipe flushing - This is done at two-yearly intervals. The water for this purpose is drawn from the system.

Pipe replacement – The pipe is normally replaced in three instances:

1) when there is frequent leakage at the same segment,
2) when the route has to be diverted,
3) due to increase in the capacity.

System monitoring - In selected points in the system (usually the ends of the system), the pressure is monitored continuously but only during the seasonal peaks in summer. Pressures and flows are measured automatically in all the pumping facilities, throughout the whole year. All the records are kept in the head office.

Leakage - The leakage level in the system is approximately 5 % of total production. It is predominantly (by number of leaks) in the service lines. Most of the registered breakages in the distribution system are for AC pipes. No leakage detection programme exists due to the low leakage rate. Very often the consumers report the leaks. Precise data about the breaks is recorded on the computer. Leak repairs normally take 3-4 hours and this service is available 24-hours a day. The team on duty has a truck equipped with all the necessary tools. Outside regular working hours the truck is always with one of the technicians, and when required they can drive directly to the location of the failure, or to the district centre if additional information is necessary.

Metering - All the service connections in the system have water meters installed. The reading of domestic consumers' meters is carried out once a year and for large consumers four times a year. The meters are replaced every 7-10 years. The company has its own workshop for meter maintenance. On average approximately 20,000 meters are replaced or rehabilitated per year. All the records with respect to meters are computerized. The estimated meter under-registration is approximately 1.5 % of the total delivery into the system.

Training and research - There are 'on the job' and 'off the job' training programmes. A compulsory two-year training for technical staff is organised by VEWIN. The company arranges different types of short training programmes. The company does not invest substantially in Research & Development, which is carried out in cooperation with the KIWA institute (nowadays known as KWR).

References

Abu-Madi, M. and Trifunović, N. 2013. *Impacts of supply duration on the design and performance of intermittent water distribution systems in the West Bank.* Water International, DOI:10.1080/02508060.2013.794404.

Abu-Thaher. A.I.K. 1998. *Leak Detection and Control for Ramallah District, Palestine.* IHE Delft, the Netherlands. MSc EE.055. (Unpublished)

Achttienribbe, G.E., 1993. *De Nederlander en zijn waterverbruik (vervolg).* Journal H_2O 13/93. (in Dutch language).

ADB. 1997. *Second Water Utilities Data Book: Asian and Pacific Region.* Edited by McIntosh, A.C. and Yniguez, C.E. Asian Development Bank.

ADB. 2004. *Water in Asian Cities.* Asian Development Bank.

Ang, W.H., Jowitt, P. W. 2006. *Solution for Water Distribution Systems under Pressure-Deficient Conditions,* Journal of Water Resources Planning and Management, 132: p.175-182.

Arulraj, G.P., Rao, H. S. 1995. *Concept of Significance Index for Maintenance and Design of Pipe Networks,* Journal of Hydraulic Engineering, 121(11): p.833-837.

Awyaja, L.D.L.C. 2015. *Assessing and Controlling Leakages in a Selected Water Supply Scheme in Towns North of Colombo, Sri Lanka.* MSc Thesis, Asian Institute of Technology, Bangkok, and UNESCO-IHE Delft. (Unpublished)

AWWA. 1986. *Corrosion Control for Operators.* American Water Works Association.

AWWA. 2003. *Water Transmission and Distribution,* 3rd edition. American Water Works Association.

AWWA Water System Operations. 2016-1. *Water Distribution – Grades 1 & 2,* 5th edition. American Water Works Association, Association of Boards of Certification.

AWWA Water System Operations. 2016-2. *Water Distribution – Grades 3 & 4,* 5th edition. American Water Works Association, Association of Boards of Certification.

Baggelaar, P.K., Rotterdam, J.J. van, Hoven, T.J.J. van den, 1988. *Kenmerken van storingen in transport- en hoofdleidingnetten; analyse van afsluitingen.* KIWA, Nieuwegein. (in Dutch language).

Barr, D.I.H., 1975. *Two Additional Methods of Direct Solution of the Colebrook-White function.* Proceedings Institution of Civil Engineers, 59(2): p.827-835.

BGW. 2001. *Wasserstatistik, Berichtsjahr 1999.* (in German language).

Bhave, P.R. 1991. *Analysis of Flow in Water Distribution Networks.* Technomic Publishing Co. Inc.

Blokland, M.W., Trifunović, N. 1994. *Engineering Economy - Notes and Problems.* IHE-Delft, the Netherlands. (Unpublished)

Brandon, T.W. 1984. *Water Distribution Systems.* The Institution of Water Engineers and Scientists.

Brouwer, C., Heibloem, M. 1986. *Irrigation Water Needs - Training Manual No.3.* UNESCO.

Bult, E.J. 1992. *IHE - Delft bloeit op een beerput*. IHE-Delft, the Netherlands. (in Dutch language)

BWS, 2003. *Belgrade Waterworks and Sewerage Facts 2003*.

BWS, 2015. *Izveštaj o tehničkim pokazateljima efiksanosti BVK za 2014-tu godinu*. (in Serbian language)

CBS – Centraal Bureau voor Statistiek. 2015. *Population, households and population dynamics; from 1899* (in the Netherlands). (Web link: http://statline.cbs.nl/statweb/publication/, access date October 2016).

Coe, A.L. 1978, *Water Supply and Plumbing Practices in Continental Europe*. Thames Water Authority, Hutchinson Benham Ltd.

Cohen, J., Konijnenberg, W.F. 1994.*Reistijdonderzoek drinkwater in het leidingnet van GW en PWN onmisbaar*. Journal H$_2$O 24/94. (in Dutch language)

Cornish, R.J. 1939. *The Analysis of Flow in Networks of Pipes*. Journal of Institution of Civil Engineers, 13: p.147-154.

Cross, H. 1936. *Analysis of Flow in Networks of Conduits or Conductors*. Bulletin No. 286, University of Illinois, Urbana IL, USA.

Cullinane, M.J., Goodrich, J., Goulter, I.C. 1989. *Water Distribution System Evaluation*. in L.W. Mays. ed., *Reliability Analysis of Water Distribution Systems*. ASCE, New York: p.85-101.

Cvjetković, M. 2008. *Belgrade Waterworks and Sewerage – Development and Reforms from 2000 until 2008*. ISBN 978-86-87531-01-7.

De Garmo, P., Sullivan,W.G., Botandelli, J.A.,1993. *Engineering Economy*. Macmillan Publishing Co., New York

Dovey, W.J., Rogers, D.V. 1993. *The Effect of Leakage Control and Domestic Metering on Water Consumption in the Isle of Weight*. Journal IWEM, 1993 (7): p.156-161.

Ductile Iron Pipe Research Association (DIPRA). 2017. *Corrosion Control – Polyethylene Encasement* (Web link: https://www.dipra.org/ductile-iron-pipe-resources/technical-publications/corrosion-control, access date September 2019).

Duyl, L.A. van, Trifunović, N. 1993. *Applied Hydraulics in Sanitary Engineering – Lecture Notes*, IHE-Delft, the Netherlands. (Unpublished)

Engelhardt, M.O., Skipworth, P.J., Savic, D.A., Saul, A.J., and Walters, G.A. (2000). *Rehabilitation Strategies for Water Distribution Networks: a Literature Review with a UK Perspective*. Urban Water, 2(2000), 153-170.

Ent, W.I. van der, 1993. *Aleid 7.0 - User's Guide*. KIWA, Nieuwegein.

Environment Agency England and Wales (EAE&W), 2000. *Optimum Use of Water for Industry and Agriculture: Best practice manual*. Research and Development Technical Report W254

EUREAU. 2009. *EUREAU Statistics Overview on Water and Wastewater in Europe 2008*. European Federation of National Associations of Water & Wastewater Services (Web link: www.eureau.org, access date October 2016).

Farley, M., 2001. *Leakage Management and Control*. World Health Organisation, Geneva

Gabrić, S., 1996. *Pressure Related Demand Modelling and Reliability Assessment: Case Study Zadar, Croatia*. IHE-Delft, the Netherlands. MSc EE.255. (Unpublished)

Gargano, R., and Pianese, D. (2000). *Reliability as a Tool for Hydraulic Network Planning*. Journal of Hydraulic Engineering, ASCE, 126(5), 354-364.

George, A., Liu, J. W-H. 1981. *Computer Solution of Large Sparse Positive Definite Systems*, Prentice-Hall, Inc., Englewood Cliffs, NJ

Giustolisi, O., Savić, D., Kapelan, Z. 2008. *Pressure-Driven Demand and Leakage Simulation for Water Distribution Networks*, Journal of Hydraulic Engineering, ASCE, 134: p.626-635.

Goulter, I.C., and Coals, A.V. (1986). *Quantitative Approaches to Reliability Assessment in Pipe Networks*. Journal of Transportation Engineering, ASCE, 112(3), 287-301.

Gupta, R., Bhave, P.R. 1996. *Comparison Methods for Predicting Deficient- Network Performance*, Journal of Water Resources Planning and Management, 122(3): p.214-217.

Gupta, R. and Bhave, P.R. 1996. *Reliability-based Design of Water Distribution Systems*, Journal of Environmental Engineering Division, ASCE, 122(1), 51-54.

Haestad Methods, Walski T.M. Chase, D.V., Savić D.A., Grayman, W., Beckwith, S., Koelle, E. 2003. *Advanced Water Distribution Modeling and Management*. Haestad Press, USA. (Reprint Bentley Institute Press, 2007)

Haaland, S.E., 1983. *Simple and Explicit Formulas for the Friction Factor in Turbulent Flow*. Journal of Fluids Engineering, 105 (1): p.89-90.

Hemed, S.H., 1996. *Computer Modelling of Zanzibar Distribution Network*. IHE-Delft, the Netherlands. MSc EE.231. (Unpublished)

Hirner, W. 1997. *Maintenance and Rehabilitation Policies of Urban Water Distribution Systems*. Water Supply, 15 (1): p.59-66.

Holzenberger, K., Jung, K. 1990. *Centrifugal Pump - Lexicon, 3rd edition*. KSB.

Hoogsteen, K.J., 1992. *Leak Detection - Lecture Notes*. IHE-Delft, the Netherlands (unpublished).

HR Wallingford, 2003. *Handbook for the Assessment of Catchment Water Demand and Use*. From DFID sponsored project, done in cooperation with Ministry of Rural Resources and Water Development (Zimbabwe), University of Zimbabwe, University of Zambia and University of Zululand (South Africa).

Hofmann, F. 2011. *Principles of Electromagnetic Flow Measurement – 2nd Edition*. KROHNE Messtechnik GmbH, Duisburg, Germany.

Hoven, T.J.J. van den, Eekeren, M.W.M. van, 1988. *Optimal Composition of Drinking Water*. KIWA Report No.100. KIWA, Nieuwegein.

Hoven, T.J.J. van den, Kramer, J.E., Hoefnagels, F.E.T., Vreeburg, J.H.G. 1993. *Milieu-effecten van leidingmaterialen*. Journal H_2O 22/93. (in Dutch language)

Idel'cik, I.E., 1986. *Memento des pertes de charge*. Third Edition (in French language), Editions Eyrolles, Paris

Istria Tourist Board. 2015. *Tourist Arrivals and Nights by Tourist Offices* (Web link: http://www.istra.hr/en/pr/statistics/2015, access date October 2016).

IWA. 2000. *Losses from Water Supply Systems: Standard Terminology and Recommended Performance Measures*. IWA Blue Pages. International Water Association.

IWA. 2018. *Non-Revenue Water Management in Low and Middle Income Countries*. Workshop, IWA World Water Congress & Exhibition, 16-21 September, Tokyo, Japan.

Kafodya. V.J. 2010. *Water Loss Management in Blantyre, Malawi*. UNESCO-IHE, Delft, the Netherlands. MSc MWI 2010-08. (Unpublished)

K-water.2015. *Multi-regional Water Supply System*. (Web link: http://english.kwater.or.kr/eng/busi/multPage.do?s_mid=1180, access date October 2016)

KSB. 1990. *Centrifugal Pump Design*. KSB.

KSB. 1992. *Pompenkatalogus*. KSB. (in Dutch language)

Kujundzić, B.1996. *Veliki vodovodni sistemi*. Udruženje za tehnologiju vode i sanitarno inžinjerstvo, Beograd. (in Serbian Language)

Lallana, C. et al. 2001. *Sustainable Water Use in Europe, Part 2: Demand Management*. European Environment Agency (EEA) – Copenhagen, Denmark (Web link: http://www.iied.org, access date January 2005).

Lambert, A. O., Brown, T. G., Takizawa, M., and Weimer, D. 1999. *A Review of Performance Indicators for Real Losses from Water Supply Systems*. J. Water SRT – Aqua Vol. 48, No.6, p 227–237.

Lansey, K. 2012. *Sustainable, Robust, Resilient, Water Distribution Systems*. Key Note Address. In Proceedings of the 14th WDSA Conference, Adelaide, Australia.

Larock, B.E., Jepson, R.W., Watters, G.Z., 2000. *Hydraulics of Pipeline Systems*. CRC Press LLC.

Liou, C.P.L. 1998. *Limitations and Proper Use of the Hazen-Williams Equation*. Journal of Hydraulic Engineering, September 1998: p.951-954.

The LEAKSSuite Library. 2019. (Web link: https://www.leakssuitelibrary.com/ access date October 2019).

Leeflang, K.W.H. 1974. *Ons drinkwater in de stroom van de tijd*. VEWIN (in Dutch)

Legrand, L., Leroy, P. 1990. *Prevention of Corrosion and Scaling in Water Supply Systems*. Ellis Horwood Ltd.

Mays, L.W., 2000. *Water Distribution Systems Handbook*. McGraw-Hill, New York

Mc Ilroy, M.S. 1949. *Pipeline Network Flow Analysis Using Ordinary Algebra*. Journal of AWWA, 41(6): p.422-428.

The Management and Implementation Authority of the Great Man-made River Project. 1989. *The Great Man-made River Project*. G. Canale & C. S.p.A.,Turin, Italy.

McIntosh, A.C. 2003. *Asian Water Supplies – Reaching the Urban Poor*. Asian Development bank/IWA, UK.

Merks, C., Lambert, A., Trow, S. 2015 *Improved Leakage Management by European Water Utilities to Safeguard Europe's Water Resources*. Presentation at the North American Water Loss Conference 2015, Atlanta, Georgia.

Netze BW GmbH. 2010. *Kenndaten zur Stuttgarter Trinkwasserversorgung 2010*. (Web link: https://www.netze-bw.de/unternehmen/aktuelles-und-projekte/ trinkwasser-stuttgart/index.html, in German language, access date January 2017).

Neubeck, K. (2004). *Practical Reliability Analysis*. Pearson Education Inc., Ohio USA

Nixon, S.C., Lack, T.J., Hunt, D.T.E., Lallana, C., Boschet, A.F. 2000. *Sustainable Use of Europe's Water? State, Prospects and Issues*. European Environment Agency (EEA) – Copenhagen, Denmark. (Web link: http://www.eea.eu.int, access date January 2005).

Obradović, D. 1991. *Mathematical Modelling of Water Supply Systems - Lecture Notes*, IHE Delft, the Netherlands. (Unpublished)

Obradović, D., Lonsdale, P., 1998. *Public Water Supply: Models, Data and Operational Management*. E & FN Spon, London

Orgilt, A., 2008. *Optimization of valve operation Water Distribution Network: Case Study Ulaanbaatar city, Mongolia*. UNESCO-IHE, Delft, the Netherlands. MSc MWI-2008/05. (Unpublished)

Ostfeld,A. (2004). *Reliability Analysis of Water Distribution Systems*, Journal of Hydroinformatics, IWA, 6(4), 281-294.

Ozger, S.S., Mays, L.W. 2003. *A Semi-Pressure-Driven Approach to Reliability Assessment of Water Distribution Networks*. In Proceedings of the 30th IAHR Congress, p.345-352, Thessaloniki, Greece.

Prasad, T. D., and Park, N.S. (2004). *Multiobjective genetic algorithms for design of water distribution networks*. J. Water Resources Planning and Management. ASCE, 130(1), p 73–82.

Pieterse, L.J. 1991. *Project Management Guidelines - Lecture Notes*. IHE-Delft, the Netherlands. (Unpublished)

Planning Commission, Government of India. 2013. *Twelfth Five Year Plan (2012/2017)*. SAGE Publications India Pvt Ltd. New Delhi, India

Piller, O., Zyl, J.E. van, 2009. *Pressure-Driven Analysis of Network Sections Supplied via High-Lying Nodes*. In Proceedings of the 10th International Conference on Computing and Control for the Water Industry, CCWI 2009 - Integrating Water Systems, p.257-262, 1-3 September, Sheffield, UK.

Pont-A-Mousson. 1992. *A Comprehensive Ductile Iron System*. Pont-A-Mousson.

Radojković, M., Klem, N., 1989. *Primena računara u hidraulici*. IRO Građevinska knjiga, Beograd. (in Serbian Language)

Rico-Amoros, A.M., Olcina-Cantos, J., Sauri, D., 2009. *Tourist Land Use Patterns and Water Demand: Evidence from the Western Mediterranean*. Journal of Land Use Policy, 26(2009): p.493-501.

Roberson, J.A., Crowe, C.T., 1995. *Engineering Fluid Mechanics*. John Wiley & Sons, Inc.

Robinson, W.C., 1993. *Testing Soil for Corrosiveness*. Materials Performance, Vol. 32, Intermountain Corrosion Service Inc., National Association of Corrosion Engineering Publisher, Houston, Texas , p. 56-58

Rossman, L.A., 2000. *Epanet 2 Users Manual*. Water Supply and Water Resources Division, U.S. Environmental Protection Agency, Cincinnati, OH, EPA/600/R-00/057

Seidel, K., Lange, G. 2007. *Direct Current Resistivity Methods. In: Environmental Geology. Handbook of Field Methods and Case Studies*. (Eds.) Knödel, K., Lange, G., Voigt, H.J. Springer Verlag, Berlin, Germany

Sextus Julius Frontinus. *De Aquaeductu (The Aqueducts of Rome* - translated by Bennett C.E.). 1925. Loeb, USA

Shamir, U. and Howard, C.D.D. (1979). *An Analytical Approach to Scheduling Pipe Replacement*. Journal of AWWA, Vol.71, No.5, pp. 248-258.

Shinstine,D.S., Ahmed,I., and Lansey,K. (2002). *Reliability/availability analysis of municipal water distribution networks: Case studies*. Journal of Water Resources Planning and Management, ASCE,128(2),140-151.

Slaats, P.G.G., Meerkerk, M.A., Hofman-Caris, C.H.M. 2013. *Conditionering: de optimale samenstelling van drinkwater; Kiwa-Mededeling 100 – Update 2013*. KWR 2013.069. KWR, Nieuwegein. (in Dutch language)

Smet, J., van Wijk, C. (eds.), 2002. *Small Community Water Supplies: Technology, People and Partnership*. IRC International Water and Sanitation Centre. Delft, the Netherlands.

Smith, L.A., Fields, K.A., Chen, A.S.C., Tafuri, A.N., 2000. *Options for Leak and Break Detection and Repair for Drinking Water Systems*. Battelle Press, Columbus, Ohio.

Snoeyink, V.L. 2002. *Water Quality Changes in Distribution Systems*. The Association of Environmental Engineering and Science Professors Lecture, American Water Works Association (AWWA) Annual Conference, New Orleans, LA, June 2002.

Sothern African Development Community (SADC) 1999. *Integrated Water Resources Development and Management in the Southern African Development Community Conference*

Sophocleous, S. 2018. *Development of the Next Generation of Water Distribution Network Modelling Tools Using Inverse Methods*. PhD Dissertation, University of Exeter, UK.

Streeter, W.L., Wylie, E.B., 1985. *Fluid Mechanics*. Eighth Edition. McGraw-Hill, Inc.

Su, Y.C., Mays, L.W., Duan, N. and Lansey, K. (1987). *Reliability-Based Optimization Model for Water Distribution Systems*. Journal of Hydraulic Engineering Division, ASCE, 114(12), 1539-1556.

Swamee, P.K., Jain, A.K., 1976. *Explicit equations for pipe-flow problems*. Journal of the Hydraulics Division, 102(5): p.657-664.

Tanyimboh, T. T., Tabesh, M., and Burrows, R. 2001. *Appraisal of Source Head Methods for Calculating Reliability of Water Distribution Networks*, J. Water Resources Planning and Management, ASCE, 127(4), 206-213.

Thiadens, A., 1996. *Unaccounted-For Water – Handout for short course on 'Low-Cost Water Supply and Sanitation'*. IHE-Delft, the Netherlands. (Unpublished)

Thiel, L. van, 2014. *Watergebruik Thius 2013. Rapport in opdracht G5707 van VEWIN*. TNS Nipo, Amsterdam (in Dutch language).

Thornton, J., 2002. *Water Loss Control Manual*. McGraw-Hill.

Tobias, P.A., and Trindade, D.C. 1995. *Applied reliability*. Second Edition, Chapman & Hall/CRC, New York, NY.

Todini, E., Pilati, S. 1987. *A Gradient Method for the Analysis of Pipe Networks.* International Conference on Computer Applications for Water Supply and Distribution, Leicester Polytechnic, UK

Todini, E., Pilati, S. 1988. *A Gradient Algorithm for the Analysis of Pipe Networks.* in B. Coulbeck and C.Orr. eds., *Computer Applications in Water Supply, Volume 1 – Systems Analysis and Simulation.* Research Studies Press Ltd., John Wiley & Sons Inc., New York: p.1-20.

Todini, E. 2000. *Looped water distribution networks design using a resilience index based heuristic approach.* Urban Water, Elsevier, Article No 63, p 1–8.

Todini, E. 2006. *Towards Realistic Extended Period Simulations EPS in Looped Pipe Network,* In Proceedings of the 8th Annual International Symposium on Water Distribution System Analysis, ASCE, Reston, USA. (CD-ROM)

Trevor Hodge, A., 1992. *Roman Aqueducts & Water Supply.* Gerald Duckworth & Co. Ltd. UK

Trifunović, N., Abu-Madi, M.O. R. 1999 *Demand Modelling of Distribution Systems With Individual Storage.* Proceedings from the 26th A.S.C.E. Annual Water Resources Planning & Management Conference, Tempe, Arizona.

Trifunović, N., Blokland, M.W. 1993. *Computer Calculation of Hodeidah Water Distribution System.* IHE-Delft, the Netherlands. (Unpublished)

Trifunović, N. 2012. *Pattern Recognition for Reliability Assessment of Water Distribution Networks.* PhD Dissertation, TU-Delft, the Netherlands.

Tung, Y.K. 1996. *Uncertainty Analysis in Water Resources Engineering.* Stochastic Hydraulics '96 Editor: Goulter, I. and Tickle, K., A.A. Balkema Publishers, p29-46.

UK Water Industry. 2015. *Guide to Pressure Testing of Pressure Pipes and Fittings for Use by Public Water Suppliers.* Water Industry Information and Guidance Note IGN 4-01-03.

UN. 2012. *World Urbanization Prospects – The 2011 Revision.* United Nations Department of Economic and Social Affairs. New York, USA

VEWIN. *Waterleidingstatistiek 1960-2001.* Vereniging van waterbedrijven in Nederland (in Dutch language)

VEWIN. 2014. *Synopsis Watergebruik Thuis 2013.* Vereniging van waterbedrijven in Nederland (Web link: www.vewin.nl, in Dutch language, access date October 2016).

VEWIN. 2015. *Dutch Drinking Water Statistics 2015.* VEWIN Association of Dutch water companies (Web link: www.vewin.nl, access date October 2016).

Vreeburg, J.H.G., 2007. *Discolouration in Drinking Water System: A Particular Approach.* PhD Thesis, TU-Delft.

Vreeburg, J.H.G., Hoven, T.J.J. van den, 1994. *Maintenance and Rehabilitation of Distribution Networks in the Netherlands.* KIWA, Nieuwegein.

Vreeburg, J.H.G., Hoven, T.J.J. van den, Hoogsteen, K.J. 1993. *A Quantitative Method to Determine Reliability of Water Supply Systems.* Proceedings from the 19th IWSA Congress, Budapest.

Ward, D., McKague, K., 2007. *Water Requirements of Livestock.* Fact Sheet Order No. 07-023, AGDEX 716/400, Ontario Ministry of Agriculture, Food and Rural Affairs.

Watson, T., Christian, C., Mason, A., and Smith, M. 2001. *Maintenance of Water Distribution Systems.* Proceedings from the 36th Annual Conference of the Operational Research Society of New Zealand, University of Canterbury, 57-66.

WB. 1996. *Water and Wastewater Utility Data,* 2nd edition. World Bank.

Wessex Water plc. 1993. *Wesnet - Technical Reference Guide.* Oakdale Printing Co. Ltd. Poole.

West, J.M., 1980. *Basic Corrosion and Oxidation.* Ellis Horwood Ltd. Chichester.

Weimer, D., 1992. *Leakage Control,* Water Supply, Volume 10, p.169-176.

WHO, UNICEF, WSSCC. 2000. *Global Water Supply and Sanitation Assessment 2000 Report*. World Health Organisation, United Nations Children's Fund, Water Supply and Sanitation Collaborative Council.

WHO. 2004. *Guidelines for Drinking-water Quality*, 3rd edition. World Health Organisation.

WHO. 2004. *Safe Piped Water: Managing Microbial Water Quality in Piped Distribution Systems,* edited by Richard Ainsworth. Published by IWA.

Williams, R.B., Culp, G.L., 1986. *Handbook of Public Water System.* Van Nostrand Reinhold Company Inc., New York.

Wood, D.J., Charles, C.O.A. 1972. *Hydraulic Network Analysis Using Linear Theory.* Journal of Hydraulic Division of ASCE, 98(HY76): p.1157-1170.

World Climate Guide. 2016. *Porec Climate Guide, Istria and Kvarner Riviera, Croatia* (Web link: http://www.worldclimateguide.co.uk/climateguides, access date January 2017).

Wu, Z.Y., Wang, R.H., Walski, T.M., Yang, S.Y., Bowdler, D., Baggett, C.C. 2006. *Efficient Pressure-Dependant Demand Model for Large Water Distribution System Analysis,* In Proceedings of the 8th Annual International Symposium on Water Distribution System Analysis, ASCE, Reston, USA. (CD-ROM)

Wu, Z.Y., Wang, R.H., Walski, T.M., Yang, S.Y., Bowdler, D., Baggett, C.C. 2009. *Extended Global-Gradient Algorithm for Pressure-Dependant Water Distribution Analysis,* Journal of Water Resources Planning and Management, 135: p.13-22.

Wu, Z.Y., Farley, M., Turtle, D., Kapelan, Z., Boxall, J., Mounce, S., Dahasahasra, S., Mulay, M., Kleiner, Y. 2011. *Water Loss Reduction.* Bentley Institute Press, Exton, Pennsylvania.

WWAP (United Nations World Water Assessment Programme). 2015. *The United Nations World Water Development Report 2015: Water for a Sustainable World.* Paris, UNESCO.

Zhou, Y. 2018. *Deterioration and Optimal Rehabilitation Modelling for Urban Water Distribution Systems*. PhD Dissertation, TU-Delft, the Netherlands.

Zyl, J.E. van, 2011. *Introduction to Integrated Water Management - Edition 1*, Water Research Commission TT 490/11, the Republic of South Africa.

Zwan, J.T. van der, Blokland, M.W. 1989. *Water Transport and Distribution - Lecture Notes, Part 1*, IHE-Delft, the Netherlands. (Unpublished)

Index